A Bibliographic Guide to North American Industry

History, Health, and Hazardous Waste

Dale A. Stirling

The Scarecrow Press, Inc.
Lanham, Maryland • Toronto • Plymouth, UK
2009

SCARECROW PRESS, INC.

Published in the United States of America
by Scarecrow Press, Inc.
A wholly owned subsidiary of
The Rowman & Littlefield Publishing Group, Inc.
4501 Forbes Boulevard, Suite 200, Lanham, Maryland 20706
www.scarecrowpress.com

Estover Road
Plymouth PL6 7PY
United Kingdom

British Library Cataloguing in Publication Information Available

Library of Congress Cataloging-in-Publication Data

Stirling, Dale A.
A bibliographic guide to North American Industry : history, health, and hazardous waste / Dale A. Stirling.
p. cm.
Includes bibliographical references and indexes.
ISBN-13: 978-0-8108-6701-7 (cloth : alk. paper)
ISBN-10: 0-8108-6701-X (cloth : alk. paper)
ISBN-13: 978-0-8108-6702-4 (ebook)
ISBN-10: 0-8108-6702-8 (ebook)
1. Industries–Environmental aspects–North America–Bibliography. 2. Pollution prevention–North America–Bibliography. I. Title. II. Title: Guide to North American Industry : history, health, and hazardous waste. III. Title: North American Industry : history, health, and hazardous waste.
Z5863.P7S85 2009
[HC95.Z9 E5]
016.36373'1–dc22 2008042140

∞™ The paper used in this publication meets the minimum requirements of American National Standard for Information Sciences—Permanence of Paper for Printed Library Materials, ANSI/NISO Z39.48-1992.
Manufactured in the United States of America.

Contents

Foreword

I am a historian. My clients are stumped. They bought a large plot of raw land, and now that they have started to prepare it for development, the graders have uncovered a trough of some particularly virulent organic chemical wastes. How could this be? The bank had required a phase one environmental site assessment, which a competent engineering firm had conducted, all in accordance with ASTM due diligence standards. Sure, there had been a paper mill operating on the adjacent property during the 1940s, but could that have generated these wastes and left such a mess to find so many decades later? How do I, as a professional historian, help them figure this out?

Now that this bibliography sits on my shelves, the very first thing I do is take it down and thumb through the "Manufacturing" section. There, I find that section 322 is devoted to paper manufacturing. Next, referring under this heading to "Readings in Waste Management and Pollution Prevention," I find what I need: H. J. Skinner, "Waste Problems in the Pulp and Paper Industry," published in a trade journal in 1939! Now, if I had known that H. J. Skinner had written such an article, I could have plugged his name, along with some keywords, such as "paper," and "waste," and "industry," into Google Scholar and received the full citation. However, I did not know that, and searching with those keywords alone yields 262,000 results, in no particular order of usefulness. If I add "1940s" to narrow down the relevant date range, I reduce the number of hits to 15,700, none of which, as far as I can tell, includes the Skinner essay, because its publication date was 1939. There are things that we cannot yet do on the Internet, and identifying targeted readings, screened for their utility by an expert in the field, is one of them. Dale A. Stirling, a historian consulting in the practical matters of environmental history for more than two decades, is such an expert. This volume is a product of his research, experience, and devotion, and I, for one, expect to dog-ear mine in short order.

Shelley Bookspan
Santa Barbara

Preface

The purpose of this reference work is to provide a starting point for those responsible for researching historical aspects of North American industry. It is intended to help answer questions pertaining to the historical development of industry and its subsequent impacts on human health and the generation of hazardous waste.

Those who may find this reference work useful include students of industrial evolution, industrial hygienists, doctors of occupational and environmental medicine, attorneys specializing in toxic tort litigation, and researchers and historians who try to answer the difficult questions of who could have, should have, and did know about worker health impacts and hazardous waste generation.

The reader may find this reference work different from others in that many citations are not the most current. The reason for this is that this work assumes that every era of information about industrial processes and subsequent health and hazardous waste impacts is pertinent. This is especially critical in litigation that attempts to answer historical questions of worker health and pollution or where the questions "Who could or should have known?" "When should they have known?" or "Why didn't they know?" are asked.

Acknowledgments

I thank those who have instilled in me a love of the past. Those who cajoled, pushed, challenged, and supported this include my wonderful parents, my wife Stephanie, and Dr. Stephen Haycox of the University of Alaska, Anchorage, who was the first to challenge me intellectually.

Introduction

For nearly 30 years, I have conducted research into the environmental practices of North American industry and commerce. During this time, I have often been asked to determine the history of an industry, the health and safety issues related to an industry, and how an industry managed its hazardous waste and chemicals of concern.

These questions were asked primarily because individuals and corporations needed to know the history of industrial practices in order to determine the environmental liability of real property they owned, leased, rented, or planned to purchase or sell. In addition, individuals and corporations needed to know what the potential human health impacts were of exposure to industrial waste streams in response to litigation or impending regulatory changes. Unfortunately, there was no convenient "guide" to this information—thus my forays into this reference work beginning in 2001. At about the same time I was conducting research based on these questions, I found that the North American Industry Classification System (NAICS) provided a convenient framework for identifying virtually every type of industry. Therefore I used the NAICS framework to assist me in compiling this bibliographic guide to North American industry. The guide includes industries commonly known to generate waste streams of concern to the environment and human health, as well as industries involved in planning, writing, imagining, or proposing, or that are otherwise based on administrative, economic, intellectual, and legal activities—pursuits not commonly known to generate waste streams of concern to the environment and human health. As those industries are part of the NAICS framework, suggested readings in history for those industries are included in this guide, but for the most part there are few or no corresponding readings in health and safety or in hazardous waste management and pollution prevention. The primary reason for this is that there simply has not been research into matters of waste management, pollution prevention, and heath impacts for those industries. However, because all industries have been subjected to historical study, they are included in this guide.

SELECTION OF BIBLIOGRAPHIC ENTRIES

Focused literature searches were conducted using standard reference databases such as DIALOG databanks, the National Library of Medicine's PUBMED, and ABC-CLIO's American History & Life and Historical Abstracts. The key criteria for choosing a bibliographic entry were, first, its direct relationship to the industry (whether history, waste stream, or health related); second, whether its abstract and/or text provided clear evidence that the information provided by the author offered enough information to be useful to the reader of this reference work; and last, it may have been a resource I utilized in my research. Accordingly, references included in this work range from peer-reviewed journal articles to books, from dissertations to government reports. Although the NAICS framework is specific to the United States, Canada, and Mexico, to rein in the length of this reference work, bibliographic citations are primarily related to industrial activity in the United States. However, significant research into worker health and safety and hazardous waste management and pollution prevention has been conducted in other countries (notably Europe), and therefore information that is applicable to U.S. industrial settings is also included in this guide.

READINGS IN HISTORY

Suggested readings in history consist primarily of journal articles, books, and dissertations. Readings were selected based on in-depth literature searches and subsequent review of abstracts. The intent of these readings is to provide the reader with sources of information on the history of the industries outlined in the NAICS.

READINGS IN HAZARDOUS WASTE MANAGEMENT AND POLLUTION PREVENTION

Suggested readings in hazardous waste management and pollution prevention consist primarily of journal articles, books, and professional reports. They were selected based on in-depth literature searches and subsequent reviews of abstracts. The intent of these suggested readings is to provide the reader with sources of information on hazardous wastes generated by industry and approaches to managing that waste and preventing further pollution.

READINGS IN HEALTH AND SAFETY

Suggested readings in health and safety consist primarily of journal articles and books. They were selected based on in-depth literature searches and subsequent review of abstracts. The intent of these readings is to provide the reader with sources of information on the chief health and safety issues facing industry. Emphasis is placed on literature relating to specific industrial practices that are known to present risks to human health, as well as specific chemicals commonly known to present human health risks (e.g., lead, mercury, chromium, etc.).

Sector 11: Agriculture, Forestry, Fishing, and Hunting

111: Crop Production
1111: Oilseed and Grain Farming
1112: Vegetable and Melon Farming
1113: Fruit and Tree Nut Farming
1114: Greenhouse, Nursery, and Floriculture Production
1119: Other Crop Farming

READINGS IN HISTORY

1. Bachus, E. J. 1981. Who Took the Oranges Out of Orange County? *Southern California Quarterly* 63(2): 157–173.

2. Bennett-Kimble, L. 2000. Sweet Smell of Success. *Michigan History Magazine* 84(2): 40–45.

3. Bingen, J. and Siyengo, A. 2002. Standards and Corporate Reconstruction in the Michigan Dry Bean Industry. *Agriculture & Human Values* 19(4): 311–323.

4. Bishop, J. C. and Manisha, P. 1991. Potato Chip Industry: Cottage to Factory. *Essays in Economic & Business History* 9: 183–203.

5. Blackstoc, J. R. 1990. Laura Scudder: Southern California's Potato Chip Pioneer. *Californians* 7(5): 30–39.

6. Bogue, A. G. 1963. *From Prairie to Corn Belt: Farming on the Illinois and Iowa Prairies in the 19th Century*. Ames: Iowa State University Press.

7. Britz, B. S. 1973. Lyndhurst Greenhouse: Emblem of a Grand Society. *Historic Preservation* 25(1): 15–21.

8. Burrows, G. and Shlomowitz, R. 1992. The Lag in the Mechanization of the Sugarcane Harvest: Some Comparative Perspectives. *Agricultural History* 66(3): 61–75.

9. Carney, J. 2003. The African Antecedents of Uncle Ben in U.S. Rice History. *Journal of Historical Geography* 29(1): 1–21.

10. Coclanis, P. A. and Marlow, J. C. 1998. Inland Rice Production in the South Atlantic States: A Picture in Black and White. *Agricultural History* 72(2): 197–212.

11. Conlin, J. R. 1986. *Bacon, Beans and Galantines: Food and Foodways on the Western Mining Frontier*. Reno: University of Nevada Press.

12. Daynes, G. and Kimball, R. I. 2001. By Their Fruits Ye Shall Know Them: A Cultural History of Orchard Life in Utah Valley. *Utah Historical Quarterly* 69(3): 215–231.

13. Dethloff, H. C. 1988. *A History of the American Rice Industry, 1685–1985*. College Station: Texas A&M University Press.

14. Dimitri, C. 2002. Contract Evolution and Institutional Innovation: Marketing Pacific-Grown Apples from 1890–1930. *Journal of Economic History* 62(1): 189–212.

15. Fahey, J. 1993. Irrigation, Apples, and the Spokane Country. *Pacific Northwest Quarterly* 84(1): 7–18.

16. Ferleger, L. 1984. Cutting the Cane: Harvesting the Louisiana Sugar Cane Industry. *Southern Studies* 23(1): 42–59.

17. Fine, G. A. 1987. Community and Boundary: Personal Experience Stories of Mushroom Collectors. *Journal of Folklore Research* 24(3): 223–240.

18. Fite, G. C. 1980. Mechanization of Cotton Production Since World War II. *Agricultural History* 54(1): 190–207.

19. Fitzgerald, D. 1993. Farmers Deskilled: Hybrid Corn and Farmer's Work. *Technology and Culture* 34(2): 324–343.

20. Fitzgerald, D. K. 1990. *The Business of Breeding: Hybrid Corn in Illinois, 1890–1940*. Ithaca, NY: Cornell University Press.

21. Fornari, H. D. 1979. The Big Change: Cotton to Soybeans. *Agricultural History* 53(1): 245–253.

22. Gaiennie, B. S. 1984. Sugarcane Retreats South. *Louisiana History* 25(1): 76–78.

23. Glimp, H. A. 1995. Meat Goat Production and Marketing. *Journal of Animal Science* 73(1): 291–295.

24. Hall, S. M. 1991. Working the Black Patch: Tobacco Farming Traditions, 1890–1930. *Register of the Kentucky Historical Society* 89(3): 266–286.

25. Hargrave, L. 1990. A History of Wine Grapes on Long Island. *Long Island Historical Journal* 3(1): 3–16.

26. Harley, C. K. 1978. Western Settlement and the Price of Wheat, 1872–1913. *The Journal of Economic History* 38(4): 865–878.

27. Hart, J. F. 1986. Change in the Corn Belt. *Geographical Review* 76(1): 51–72.

28. Hart, J. P. 2000. New Dates from Old Collections: The Roundtop Site and Maize-Beans-Squash Agriculture in the Northeast. *North American Archaeologist* 21(1): 7–17.

29. Hart, J. P. and Scarry, C. M. 1999. The Age of Common Beans in the Northeast U.S. *American Antiquity* 64(4): 653–658.

30. Hayter, D. 1988. Expanding the Cotton Kingdom. *Agricultural History* 62(2): 225–233.

31. Hewes, L. 1965. Causes of Wheat Failure in the Dry Farming Region, Central Great Plains, 1939–1957. *Economic Geography* 41(4): 313–330.

32. Hiers, C. 1983. Pioneers on the Citrus Frontier. *Southern Exposure* 11(6): 20–21.

33. Hilliard, S. B. 1979. Site Characteristics and Spatial Stability of the Louisiana Sugarcane Industry. *Agricultural History* 53(1): 254–269.

34. Hudson, J. C. 1994. *Making the Corn Belt: A Geographical History of Middle Western Agriculture*. Bloomington: Indiana University Press.

35. Hurt, R. D. 1982. The Vineyards of Ohio, 1823–1900. *Northwest Ohio Quarterly* 55(1): 3–16.

36. Jett, S. C. 1977. History of Fruit Tree Raising among the Navajo. *Agricultural History* 51(4): 681–701.

37. Keller, K. W. 1995. Merchandising Nature: The H.J. Weber & Sons Nursery. *Missouri Historical Review* 89(3): 307–326.

38. Lauck, J. 2000. *American Agriculture and the Problem of Monopoly: The Political Economy of Grain Belt Farming, 1953–1980*. Lincoln: University of Nebraska Press.

39. Lee, C. M. 1996. The American Rice Industry's Organization for a Domestic Market: The Associated Rice Millers of America. *Louisiana History* 37(2): 187–199.

40. Lockeretz, W. and Wernick S. 1980. Commercial Organic Farming in the Corn Belt in Comparison to Conventional Practices. *Rural Sociology* 45(4): 708–722.

41. Macinko, G. 1985. The Ebb and Flow of Wheat Farming in the Big Bend, Washington. *Agricultural History* 59(2): 21–228.

42. MacLean, J. T. 1980. Nursery and Seed Trade Catalogs. *Journal of NAL Associates* 5(3–4): 88–92.

43. McBane, M. 2001. The House That Lemons Built: Race, Ethnicity, Gender, Citizenship and the Creation of a Citrus Empire, 1893–1919. Dissertation, University of California, Los Angeles.

44. McClintock, T. C. 1967. Henderson Luelling, Seth Lewelling and the Birth of the Pacific Coast Fruit Industry. *Oregon Historical Quarterly* 68(2): 153–174.

45. McGregor, A. C. and Shepherd, J. F. 2000. Fertilization Practices in Pacific Northwest Wheat Producing Areas. *Agricultural History* 74(2): 433–450.

46. McNeil, W. H. 1999. How the Potato Changed the World's History. *Social Research* 66(1): 67–83.

47. Miller, C. J. Tobacco Manufacturing. In *Manufacturing: A Historiographical and Bibliographical Guide,* ed. D. O. Whitten, 65–86. Westport, CT: Greenwood Press.

48. Miner, C. 1998. *Harvesting the High Plains: John Kriss and the Business of Wheat Farming, 1920–1950.* Lawrence: University Press of Kansas.

49. ——. 2000. The Wheat Empire of R. H. Garvey, 1930–1959. *Kansas History* 23(1–2): 88–99.

50. Morgan, K. 1995. The Organization of the Colonial American Rice Trade. *William & Mary Quarterly* 52(3): 433–452.

51. Morrow, H. C. 1994. Valley Vineyards: A Rich Historical Heritage. *Password* 39(2): 55–64.

52. Muehl, S. 1997. The Wild Missouri Grape and 19th Century Viticulture. *Missouri Historical Review* 91(4): 373–384.

53. Nonnecke, L. 1989. *Vegetable Production.* New York: Springer.

54. Norris, J. 2002. Bargaining for Beets: Migrants and Growers in the Red River Valley. *Minnesota History* 58(4): 196–209.

55. Olmstead, A. L. and Rhode, P. W. 2002. The Red Queen and the Hard Reds: Productivity Growth in American Wheat, 1800–1940. *Journal of Economic History* 62: 929–966.

56. Parker, W. N. 1979. Labor Productivity in Cotton Farming: The History of a Research. *Agricultural History* 53(1): 228–244.

57. Parry, B. E. 1980. Soybeans as an Indicator of the Progress of Southern Agriculture Since the Second World War. Dissertation, American University.

58. Peters, G. L. 1984. Trends in California Viticulture. *Geographical Review* 74(4): 455–467.

59. Peterson, W. J. 1968. Strawberry Time. *Palimpsest* 49(3): 81–104.

60. Pezzoni, J. D. 2001. The Architecture of Tobacco Manufacturing in 19th Century Virginia. *Pioneer America Society Transactions* 24: 69–81.

61. Pillay, T. V. R. and Kutty, M. N. 2005. *Aquaculture: Principles and Practices*. Oxford, UK: Blackwell Publishing.

62. Platt, C. V. 1991. On Branch and Vine: Ohio's Fruits and Nuts. *Timeline* 8(4): 34–41.

63. Pratt, M. J. 1993. The Strawberry King of Bradford. *Vermont History News* 44(4): 55–60.

64. Preston, J. R. 2002. Heward G. Leavitt's Influence on Sugar Beets and Irrigation in Nebraska. *Agricultural History* 76(2): 381–392.

65. Reidy, J. P. 1998. Mules and Machines and Men: Field Labor on Louisiana Sugar Plantations, 1887–1915. *Agricultural History* 72(2): 183–196.

66. Renner, G. K. 1969. Strawberry Culture in Southwest Missouri. *Missouri Historical Review* 64(1): 18–40.

67. Romero, G. 1984. *The Louisiana Strawberry Story*. Natchitoches: Northwestern State University Press.

68. Rucker, B. R. 1996. Satsumaland: A History of Citrus Culture in West Florida. *Gulf Coast History* 12(1): 60–77.

69. Rutger, J. N. 2002. Rice in the West. *Journal of the West* 41(1): 37–40.

70. Schmalz, C. L. 1989. Sugar Beets in Cache Valley: An Amalgamation of Agriculture and Industry. *Utah Historical Quarterly* 57(4): 370–388.

71. Schmitz, A. and Gray, R. The Divergence in Canada—U.S. Grain and Oilseed Policies. *Canadian Journal of Agricultural Economics* 49(4): 459.

72. Schoenman, T. 1983. Agoston Haraszthy, the Father of California Viticulture. *New Hungarian Quarterly* 24(89): 141–146.

73. Sheef, R. F. 1994. Prohibition Vineyards: The Italian Contribution to Viticulture in Missouri. *Missouri Historical Review* 88(3): 279–300.

74. Sizer, R. and Silga, W. 1981. Fruits in Iowa: A Brief History. *Palimpsest* 62(3): 80–88.

75. Smith, A. F. 2002. *Peanuts: The Illustrious History of the Goober Pea*. Urbana: University of Illinois Press.

76. Stickney, R. R. 1996. *Aquaculture of the United States: A Historical Survey*. New York: John Wiley & Sons.

77. Thu, K. M. and Durrenberger, E. P. 1998. *Pigs, Profits and Rural Communities.* Albany: State University of New York Press.

78. VanWagoner, R. S. 1991. The Lehi Sugar Factory: 100 Years in Retrospect. *Utah Historical Quarterly* 59(2): 189–204.

79. VanWilligen, J. and Eastwood, S. C. 1998. *Tobacco Culture: Farming Kentucky's Burley Belt.* Lexington: University Press of Kentucky.

80. Vaught, D. 2002. Putah Creek: Water, Land and Wheat and Community in the Sacramento Valley in the 1850s. *Agricultural History* 76(2): 326–337.

81. Verano, L. F. 1978. A Fruit for All Seasons: The Pear's Long Journey West. *Journal of the West* 17(2): 94–104.

82. Walters, W. D. 1997. *The Heart of the Corn Belt: An Illustrated History of Corn Farming in McLean County.* Bloomington, IL: McLean County Historical Society.

83. Ward, T. 1995. Farming Technology and Crop Area on Early Prairie Farms. *Prairie Forum* 20(1): 19–36.

84. Widmer, R. E. 1997. *A History of Minnesota Floriculture*. St. Paul: Minnesota Agricultural Experiment Station.

85. Wells, M. J. 2000. Politics, Locality, and Economic Restructuring: California's Central Coast Strawberry Industry in the Post–World War II Period. *Economic Geography* 76(1): 28–49.

86. Whitley, E. C. 1987. Big Hitch Wheat Farming in Eastern Washington: A Personal Account. *Pacific Northwest Quarterly* 78(1–2): 10–16.

87. Williams, J. 1996. Potatoes: A Washington Tradition. *Columbia* 10(3): 20–23.

88. Wilson, M. T. 1959. Americans Learn to Grow the Irish Potato. *New England Quarterly* 32(3): 333–350.

89. Yale, A. R. 1991. *While the Sun Shines: Making Hay in Vermont, 1789–1990*. Montpelier: Vermont Historical Society.

90. Young, J. A. and Clements, C. D. 2003. Horse Drawn Hay Making in the West. *Journal of the West* 42(1): 60–66.

91. Young, T. 1998. From Manure to Steam: The Transformation of Greenhouse Heating in the United States, 1870–1900. *Agricultural History* 72(3): 574–596.

READINGS IN HAZARDOUS WASTE MANAGEMENT AND POLLUTION PREVENTION

92. Agnew, R. H., et al. 2005. Waste Stream Utilization for Sustainable Viticulture. *Water Science Technology* 51(1): 1–8.

93. Alva, A. K., et al. 2003. Best Nitrogen and Irrigation Management Practices for Citrus Production in Sandy Soils. *Water, Air & Soil Pollution* 143(1–4): 139–154.

94. Atkins, P. F. and Sproul, O. J. 1966. Feasibility of Biological Treatment of Potato Processing Wastes. *Journal of the Water Pollution Control Federation* 38(8): 1287–1297.

95. Bachman, J. 2004. *Oilseed Processing for Small-Scale Producers*. Fayetteville, AR: National Center for Appropriate Technology.

96. Balasubramanian, V. Farmer Adoption of Improved Nitrogen Management Technologies in Rice Farming. *Nutrient Cycling in Agro EcoSystems* 53(1): 1385–1314.

97. Beck, C., et al. 2005. Activated Sludge Wastewater Treatment Plants Optimization to Face Pollution Overloads during Grape Harvesting Periods. *Water Science Technology* 51(1): 81–88.

98. Becker, K. W. 1972. Control of Pollutants for the Oilseed Industry. *Journal of the American Oil Chemistry Society* 49(2): 40A passim.

99. Bertran, E., et al. 2004. Composting Winery Waste: Sludges and Grape Stalks. *Bioresource Technology* 95(2): 203–208.

100. Boopathy, R., et al. 2001. Microbial Decomposition of Post-harvest Sugarcane Residue. *Bioresource Technology* 79(1): 29–33.

101. Bradhsaw, L., et al. 2002. Potato Byproduct Feeding in the Pacific Northwest. *Veterinary Clinics of North America: Food Animal Practice* 18(2): 339–347.

102. De Cockborne, A. M., et al. 1999. Environmental Consequences of Apple Waste Deposition in Soil. *Journal of Environmental Quality.*

103. DeVletter, R. 1972. Measures against Water Pollution in Beet Sugar Processing Industries. *Pure Applied Chemistry* 29(1): 113–128.

104. Esvelt, L. A. and Hart, H. H. 1970. Treatment of Fruit Processing Waste by Aeration. *Journal of the Water Pollution Control Federation* 42(7): 1305–1326.

105. Favretto, L., et al. 1975. Pollution of Vineyards by Atmospheric Lead. *Journal of Science Food Agriculture* 26(7): 987–992.

106. Grames, L. M. and Kueneman, R. W. 1969. Primary Treatment of Potato Processing Wastes with Byproduct Feed Recovery. *Journal of the Water Pollution Control Federation* 41(7): 1358–1367.

107. Hatfield, J. L. and Prueger, J. H. 2001. Increasing Nitrogen Use Efficiency of Corn in Midwestern Cropping Systems. *Scientific World Journal* 21(Supp. 1): 682–690.

108. He, Z. L., et al. 2004. Transport of Heavy Metals in Surface Runoff from Vegetable and Citrus Fields. *SSSAJ* 68: 1662–1669.

109. King, A. J. 1996. Water Quality and Poultry Production. *Poultry Science* 75(7): 852–853.

110. Kroboulewsky, N., et al. 2002. Environmental Risks of Applying Sewage Sludge Compost to Vineyards. *Journal of Environmental Quality* 31(5): 1522–1527.

111. Lidon, A., et al. 1999. Comparison of Drainage Estimation Methods in Irrigated Citrus or Orchards. *Irrigation Science* 19(1): 25–36.

112. Landau, C. W., et al. 1997. Monitoring Nitrogen Pollution from Sugarcane Runoff Using 15N Analysis. *Water, Air, & Soil Pollution* 98(3–4): 389–399.

113. Nanninga, H. J. and Gottschal, J. C. 1986. Microbial Problems with Waste from Potato Starch Processing. *Microbiology Science* 3(6): 179–182.

114. Panter, K. L., et al. n.d. *Pollution Prevention in Colorado Commercial Greenhouses (XCM-206)*. Fort Collins: Colorado State University.

115. Peters, H. 1972. Measures Taken against Water Pollution in Starch and Potato Processing. *Pure Applied Chemistry* 29(1): 129–141.

116. Rao, V. R., et al. 1998. Review: Status and Future Prospects of Associative Nitrogen Fixation in Rice. *World Journal of Microbiology & Biotechnology* 14(5): 621–633.

117. Southwick, L. M., et al. 2002. Potential Influence of Sugarcane Cultivation on Estuarine Water Quality of Louisiana's Gulf Coast. *Journal of Agricultural Food Chemistry* 50(15): 4393–4399.

118. Thompson, D. N., et al. 2000. Biosurfactants from Potato Process Effluents. *Applied Biochemistry & Biotechnology* 84–86: 917–930.

119. Timsina, J. and Connor, D. J. 2001. Productivity and Management of Rice-wheat Cropping Systems: Issues and Challenges. *Field Crops Research* 69: 93–132.

120. U.S. Department of Agriculture. 1992. *Agricultural Waste Management Field Handbook.* Washington, DC: USDA.

121. U.S. Environmental Protection Agency. 2000. *Profile of the Agricultural Crop Production Industry (EPA/310/R-00/001).* Washington, DC: EPA.

122. Weber, M. M. and Bahner, M. 1999. *Pollution Prevention for the Wine Industry.* Research Triangle Park, NC: RTP.

READINGS IN HEALTH AND SAFETY

123. Bamford, M. 1985. First Aid and Fruit Farming. *Occupational Health* 37(4): 162–167.

124. Borkan, J., et al. 1995. An Outbreak of Venomous Spider Bites in a Citrus Grove. *American Journal of Tropical & Medical Hygiene* 52(3): 228–230.

125. Brett, M. M. 2003. Food Poisoning Associated with Biotoxins in Fish and Shellfish. *Current Opinion Infectious Disease* 16(5): 461–465.

126. Bruynzeel, D. P. and vanKetel, W. G. 1986. Contact Dermatitis Due to Chlorothalonil in Floriculture. *Contact Dermatitis* 14(1): 67–68.

127. Chen, Y., et al. 1991. Combined Effect of Grain Farming and Smoking on Lung Function and the Prevalence of Chronic Bronchitis. *International Journal of Epidemiology* 20(2): 416–423.

128. Cusak, P. M., et al. 2003. The Medicine and Epidemiology of Bovine Respiratory Disease in Feedlots. *Australian Veterinary Journal* 81(8): 480–487.

129. David, S. S. and Goel, K. 2001. Knowledge, Attitude and Practice of Sugarcane Crushers Towards Hand Injury Prevention Strategies in India. *Injury Prevention* 7(4): 329–330.

130. Duncan, R.G. and Griffith, J. 1985. Monitoring Study of Urinary Metabolites and Selected Symptomatology among Florida Citrus Workers. *Journal of Toxicology & Environmental Health* 16(3–4): 509–521.

131. Durborow, R. M. 1999. Health and Safety Concerns in Fisheries and Aquaculture. *Occupational Medicine* 14(2): 373–406.

132. Ewers, L. M. and Tapp, L. C. 2001. Endotoxin Exposures during Potato Processing. *Applied Occupational & Environmental Hygiene* 16(12): 1079–1087.

133. Fisher, L. 2000. Tobacco Farming and Tobacco Control in the United States. *Cancer Causes Control* 11(10): 977–979.

134. Frenich, A.G., et al. 2002. Dermal Exposure to Pesticides in Greenhouse Workers. *Environmental Monitoring & Assessment* 80(1): 51–63.

135. Fulmer, S., et al. 2002. Ergonomic Exposures in Apple Harvesting: Preliminary Observations. *American Journal of Industrial Medicine* Supp. 2: 3–9.

136. Gaugitsch, H. 2002. Experience with Environmental Issues in GM Crop Production and the Likely Future Scenarios. *Toxicology Letters* 127(1–3): 351–357.

137. Hannum, S. M. 2004. Potential Impact of Strawberries on Human Health: A Review of the Science. *Critical Reviews in Food Science & Nutrition* 44(1): 1–17.

138. Hansen, J. U. 1993. Accidents Affecting Potato Harvesters. *Ugesk Laeger* 155(39): 3131–3132.

139. Joshi, S. K. and Phil, M. 2002. Rice Field Work and the Occupational Hazards. *Occupational Medicine* 4: 111–114.

140. Kim, Y. K. and Kim, Y. Y. 2002. Spider Mite Allergy and Asthma in Fruit Growers. *Current Opinion in Allergy & Clinical Immunology* 2(2): 103–107.

141. Lane, S. R., et al. 2004. The Measurement and Health Impact of Endotoxin Contamination in Organic Dusts from Multiple Sources: Focus on the Cotton Industry. *Inhalation Toxicology* 16(4): 217–229.

142. Likas, C., et al. 2001. Noxious Gases in Greenhouses. *Annals of Agricultural & Environmental Medicine* 8(1): 99–101.

143. Manfreda, J., et al. 1986. Acute Symptoms Following Exposure to Grain Dust in Farming. *Environmental Health Perspectives* 66: 73–80.

144. Manfreda, J., et al. 1989. Chronic Respiratory Disorders Relating to Farming and Exposure to Grain Dust in a Rural Adult Community. *American Journal of Industrial Medicine* 15(1): 7–19.

145. McCurdy, S. A., et al. 1996. Respiratory Health of California Rice Farmers. *American Journal of Respiratory & Critical Care Medicine* 153(5): 1553–1559.

146. McJilton, C. E., et al. 1983. Captan Exposure in Apple Orchards. *American Industry Hygiene Association Journal* 44(3): 209–210.

147. Mestres, R., et al. 1985. Survey of Exposure to Pesticides in Greenhouses. *Bulletin of Environmental Contamination & Toxicology* 35(6): 750–756.

148. NIOSH. 1987. *Preventing Entrapment and Suffocation Caused by the Unstable Surfaces of Stored Grain and Other Materials* (Pub. 88–102). Washington, DC: NIOSH.

149. Nordby, K.C., et al. 2004. Trichothecene Mycotoxins and Their Determinants in Settled Dust Related to Grain Production. *Annals of Agricultural &. Environmental Medicine* 11(1): 75–83.

150. Onsager, J. A., et al. 1970. Residues of Aldrin, Deildrin, Chlordane, and DDT in Soil and Sugar Beets. *Journal of Economic Entomology* 63(4): 1143–1146.

151. OSHA. 1996. *Grain Handling* (OSHA 3103). Washington, DC: OSHA.

152. Parrakova, E. and Fratri, I. 1980. Soil Contamination in the Environment of Intensive Farming Units. *Agricultural Wastes* 2(3): 161–170.

153. Pennsylvania State University. 2003. *Mushroom Farm Food Safety & Security Self-Assessment*. State College, PA: College of Agricultural Sciences.

154. Pratt, D. S. 1990. Occupational Health and the Rural Worker: Agriculture, Mining, and Logging. *Journal of Rural Health* 6(4): 399–417.

155. Pugh, K. J., et al. 2000. Musculoskeletal Trauma in Tobacco Farming. *Orthopedics* 23(2): 141–143.

156. Rashid, K. A., et al. 1987. Residues and Mutagenicity of Captan Applied to Apple Trees and Potential Human Exposure. *Journal of Environmental Science & Health B* 22(1): 71–89.

157. Riordan, D. C., et al. 2001. A Study of U.S. Orchards to Identify Potential Sources of *Escherichia Coli* 0157:H7. *Journal of Food Protection* 64(9): 1320–1327.

158. Rothschild, H. and Mulvey, J. J. 1982. An Increased Risk for Lung Cancer Mortality Associated with Sugarcane Farming. *Journal of the National Cancer Institute* 68(5): 755–760.

159. Rottoli, P., et al. 2003. Gardening in Greenhouses as a Risk Factor for Silicosis. *Respiration* 70(2): 221–223.

160. Samberg, Y. and Meroz, M. 1995. Application of Disinfectants in Poultry Hatcheries. *Reviews in Science & Technology* 14(2): 365–380.

161. Senthilselvan, A., et al. 1993. Predictors of Asthma and Wheezing in Adults: Grain Farming, Sex, and Smoking. *American Review of Respiratory Disease* 148(3): 667–670.

162. Sing, R., et al. 2005. Wheat Thresher Agricultural Injuries. *Asia Pacific Journal of Public Health* 17(1): 36–39.

163. Struttmann, T. W. and Reed, D. K. 2002. Injuries to Tobacco Farmers in Kentucky. *Southern Medical Journal* 95(8): 850–856.

164. Tanaka, H., et al. 2002. Workplace-related Chronic Cough on a Mushroom Farm. *Chest* 122(3): 1080–1085.

165. Vermeulen, R., et al. 2005. A Study on Immunological Responses to Exposures Encountered in Corn Farming. *Journal of Biochemistry & Molecular Toxicology* 19(3): 172.

166. Wagner, S. E., et al. 2003. Antimony Impurity in Lead Arsenate Insecticide Enhances the Antimony Content of Old Orchard Soils. *Journal of Environmental Quality* 32(2): 736–738.

167. Wang, X. R., et al. 2003. Respiratory Symptoms and Cotton Dust Exposure. *Occupational and Environmental Medicine* 60(12): 935–941.

168. Wild, L. G. and Lehrer, S. B. 2005. Fish and Shellfish Allergy. *Current Allergy & Asthma Report* 5(1): 74–79.

169. Ying, G. G. and Williams, B. 1999. Herbicide Residues in Grapes and Wine. *Journal of Environmental Science & Health B* 34(3): 397–441.

112: Animal Production
1121: Cattle Ranching and Farming
1122: Hog and Pig Farming
1123: Poultry and Egg Production
1124: Sheep and Goat Farming
1125: Animal Aquaculture
1129: Other Animal Production

READINGS IN HISTORY

170. Aduddell, R. M. and Cain, L. P. 1990. Meat and Meat Products. In *Manufacturing: A Historiographical and Bibliographical Guide,* ed. D. O. Whitten, 27–42. Westport, CT: Greenwood Press.

171. Ahmed, Z. Z. and Sieling, M. 1987. Two Decades of Productivity Growth in Poultry Dressing and Processing. *Monthly Labor Review* 110(4): 34–39.

172. Arnould, R. J. 1971. Changing Patterns of Concentration in American Meat Packing, 1880–1963. *The Business History Review* 45(1): 18–34.

173. Azzam, A. M. 1998. Competition in the US Meatpacking Industry: Is It History? *Agricultural Economics* 18(2): 107–126.

174. Barnett, D. E. 1987. Angora Goats in Texas: Agricultural Innovation on the Edwards Plateau, 1858–1900. *Southwestern Historical Quarterly* 90(4): 347–372.

175. Basset, C. A. 1983. Spring Sheep Drive on Arizona Trails. *American West* 20(3): 22–28.

176. Becker, R. B. 1973. *Dairy Cattle Breeds: Origins and Development*. Gainesville: University of Florida Press.

177. Bellush, J. 1962. Milk Price Control: History of Its Adoption, 1933. *New York History* 43(1): 70–104.

178. Boyd, W. 2001. Making Meat: Science, Technology, and American Poultry Production. *Technology and Culture* 42(4): 631–664.

179. Broadway, M. J. 1998. Where's the Beef? The Integration of the Canadian and American Beef Packing Industries. *Prairie Forum* 23(1): 19–29.

180. Buckendorf, M. 1993. The Poultry Frontier. *Idaho Yesterdays* 37(2): 2–8.

181. Bugos, G. E. 1992. Intellectual Property Protection in the American Chicken Breeding Industry. *Business History Review* 66(1): 127–168.

182. Burcham, L. T. 1961. Cattle and Range Forage in California, 1770–1880. *Agricultural History* 35(3): 140–149.

183. Caron, D. M. 1976. The Art of Beekeeping in Maryland: Past and Present. *Journal of NAL Associates* 1(3): 37–42.

184. Cross, J. A. 2004. Expansion of Amish Dairy Farming in Wisconsin. *Journal of Cultural Geography* 21(2): 77–89.

185. Cruickshank, S. 1993. There Was a Lot of Turkeys. *Idaho Yesterdays* 37(2): 9–11.

186. Dase, A. E. 2002. Where's the Beef? A Historiographical Overview of Cattle Ranching in Spanish Colonial Texas. *Journal of South Texas* 15(1): 47–56.

187. DuPuis, E. M. 2002. *Nature's Perfect Food: How Milk Became America's Drink*. New York: New York University Press.

188. Dyson, L. K. 1970. The Milk Strike of 1939 and the Destruction of the Dairy Farmers Union. *New York History* 51(5): 523–543.

189. Friedberger, M. 1999. Mink and Manure: Rural Gentrification and Cattle Raising in Southeast Texas, 1945–1992. *Southwestern Historical Quarterly* 102(3): 268–293.

190. Friedberger, M. 1994. Cattlemen, Consumers and Beef. *Environmental History Review* 18(3): 37–57.

191. Gallman, R. E. 1995. Pork Production and Nutrition during the Late 19th Century: A Weighty Issue Visited Yet Again. *Agricultural History* 69(4): 592–606.

192. Gibson, A. M. 1967. Ranching on the Southern Great Plains. *Journal of the West* 6(1): 135–153.

193. Grettler, D. J. 1999. Environmental Change and Conflict over Hogs in Early 19th Century Delaware. *Journal of the Early Republic* 19(2): 197–220.

194. Igler, D. 1995. Industrial Cowboys: Corporate Ranching in Late 19th Century California. *Agricultural History* 69(2): 201–215.

195. Iverson, P. 1994. *When Indians Became Cowboys: Native Peoples and Cattle Ranching in the American* West. Norman: University of Oklahoma Press.

196. Jelinek, L. J. 1998. Property of Every Kind: Ranching and Farming during the Gold-rush Era. *California History* 77(4): 233–249.

197. Jordan, T. G. 1972. The Origin and Distribution of Open-range Cattle Ranching. *Social Science Quarterly* 53(1): 105–121.

198. ——. 1975. The Origin and Distribution of Open-range Cattle Ranching. *Social Science Quarterly* 53(1): 105–121.

199. ——. 1977. Early Northeast Texas and the Evolution of Western Ranching. *Annal of the Association of American Geographers* 67(1): 66–87.

200. ——. 1981. *Trails to Texas: Southern Roots of Western Cattle Ranching*. Lincoln: University of Nebraska Press.

201. ——. 1993. *North American Cattle Ranching Frontiers: Origins, Diffusion, and Differentiation.* Albuquerque: University of New Mexico Press.

202. Kenny, J. K. 1963. Early Sheep Ranching in Eastern Oregon. *Oregon Historical Quarterly* 64(2): 101–122.

203. Lyson, T. A. and Gillespie, G. W. 1995. Producing More Milk on Fewer Farms. *Rural Sociology* 60(3): 493–504.

204. Maly, K. and Wilcox, B. A. 2000. A Short History of Cattle and Range Management in Hawaii. *Rangelands* 22(5): 21–23.

205. Millet, D. J. 1987. Cattle and Cattlemen of Southwest Louisiana, 1860–1900. *Louisiana History* 28(3): 311–330.

206. Murray, H. T. and Murray, J. A. 1988. Herding Sheep in the Judith Basin at the Turn of the Century. *Montana* 38(2): 68–71.

207. Otto, J. S. 1986. Open Range Cattle Ranching in South Florida: An Oral History. *Tampa Bay History* 8(2): 23–35.

208. Paul, V. 1976. The Story of Cattle Ranching in Washington. *Record* 37: 5–37.

209. Perren, R. 1971. The North American Beef and Cattle Trade with Great Britain. *Economic History Review* 24(3): 430–444.

210. Persigehl, E. S. and York, J. D. 1979. Substantial Productivity Gains in the Fluid Milk Industry. *Monthly Labor Review* 102(7): 22–27.

211. Pope, L.S. 1980. Animal Science in the 20th Century. *Agricultural History* 54(1): 64–70.

212. Postel, M. 1988. A Lost Resource: Shellfish in San Francisco Bay. *California History* 67(1): 26–41.

213. Rifkin, J. 1992. *Beyond Beef: The Rise and Fall of the Cattle Culture*. New York: Dutton.

214. Roppel, P. n.d. *Alaska Salmon Hatcheries, 1891–1959*. Anchorage: Alaska Historical Commission.

215. Santos, R. L. 1994. Dairying in California through 1910. *Southern California Quarterly* 76(2): 175–194.

216. Sayre, N. 1999. The Cattle Boom in Southern Arizona. *Journal of the Southwest* 41(2): 239–271.

217. Simmons, M. 1988. The Rise of New Mexico Cattle Ranching. *Palacio* 93(3): 4–13.

218. Methurst, D. 1999. Land Degradation and the Decline of Ranching in the Sierra Nevada Foothills, California. *Land Degradation & Development* 10(2): 161–175.

219. Starrs, P. F. 1998. *Let the Cowboy Ride: Cattle Ranching in the American West*. Baltimore: Johns Hopkins University Press.

220. Swanson, J. C. and Morrow-Tesch, J. 2001. Cattle Transport: Historical, Research, and Future Perspectives. *Journal of Animal Science* 79 (E. Supp.). E102–E109.

221. Tiddens, A. P. 1989. Aquaculture in America: The Role of Science, Government and the Entrepreneur. Dissertation, University of California, Santa Barbara.

222. Vollertsen, J. J. and Rich, A. D. 1937. Technical Developments in Meat Packing. *Industrial Engineering & Chemistry* 29(5): 485–488.

223. Wallach, B. 1981. Sheep Ranching in the Dry Corner of Wyoming. *Geographical Review* 71(1): 51–63.

224. Watkins, L. H. 1969. John S. Harbison: California's First Modern Beekeeper. *Agricultural History* 43(2): 239–248.

225. Wheeler, D. L. 1992. Winter on the Cattle Range: Western Kansas, 1884–1886. *Kansas History* 15(1): 2–17.

226. Whitaker, J. W. 1975. *Feedlot Empire: Beef Cattle Feeding in Illinois and Iowa 1840–1900.* Ames: Iowa State University Press.

227. White, J. H. 1989. Home to Roost: The Story of Live Poultry Transit by Rail. *Agricultural History* 63(3): 81–94.

228. Wright, R. F. and Huck, P. 2002. Counting Cases about Milk: Our Most "Nearly Perfect" Food, 1860–1940. *Law & Society Review* 36(1): 51–111.

READINGS IN HAZARDOUS WASTE MANAGEMENT AND POLLUTION PREVENTION

229. Christensen, L. 2002. Dairy Farming and River Condition: Investigating the Sustainable Use of Water Resources in an Agricultural Area. *Water Science & Technology* 45(11): 21–28.

230. Glimp, H. A. 1995. Meat Goat Production and Marketing. *Journal of Animal Science* 73(1): 291–295.

231. Henderson, D. J. 1999. Controlling Sheep Dip Pollution. *Veterinary Record* 144(25): 708.

232. Idaho Department of Environmental Quality. 2004. *Idaho Waste Management Guidelines for Aquaculture Operations*. Twin Falls, ID: University of Idaho.

233. Kansas State University. 2000. *Environmental Regulatory Issues Affecting Meat Processing Industry.* Manhattan, KS: Kansas State University.

234. Kebus, M. J., et al. 2003. Waste Management: Aquaculture and Fisheries. *Journal of the American Veterinary Medical Association* 223(1): 56–57.

235. Kosaric, N., et al. 1981. UHT Milk: Production, Quality and Economics. *Critical Reviews in Food Science & Nutrition* 14(2): 153–199.

236. Myers, R. A., et al. 2004. Ecology, Hatcheries and Endangered Salmon. *Science* 303(5666): 1980.

237. Oleszkiewicz, J. A. and Koziarksi, S. 1981. Management and Treatment of Wastes from Large Piggeries. *Agricultural Wastes* 3(2): 123–144.

238. Park, E., et al. 2001. Pollution Prevention and Biochemical Oxygen Demand Reduction in a Squid Processing Facility. *Journal of Cleaner Production* 9: 341–349.

239. Prioli, G. 2004. Shellfish Farming: Technologies and Production. *Veterinary Research Communication 28* (Supp. 1): 51–56.

240. OSHA. 1988. *Safety & Health Guide for the Meatpacking Industry* (OSHA 3108). Washington, DC: OSHA.

241. Scheltinga, H. M. 1972. Measures Taken against Water Pollution in Dairies and Milk Processing. *Pure Applied Chemistry* 29(1): 101–111.

242. Southern Regional Aquaculture Center. 1999. *Characterization and Management of Effluents from Aquaculture Ponds in the Southeastern United States* (SRAC Publication No. 470). Stoneville, MS: SRAC.

243. The MacArthur Agro-Ecology Research Center. 2004. *Optimization of Best Management Practices for Beef Cattle Ranching in the Lake Okeechobee Basin* (SFWMD Contract C-13414). Gainesville: University of Florida.

244. Trebler, H. A. and Harding, H. G. 1947. Industrial Wastes . . . Dairy Industry. *Industrial & Engineering Chemistry* 39(5): 609–613.

245. U.S. Environmental Protection Agency. 1995. *Guidance Manual on NPDES Regulations for Concentrated Animal Feeding Operations* (EPA/833/B-95/001). Washington, DC: EPA.

246. Virtue, W. A. and Clayton, J. W. 1997. Sheep Dip Chemicals and Water Pollution. *Science of the Total Environment* 24: 194–195: 207–17.

247. Williams, A. G. and Evans, M. R. 1981. Storage of Piggery Slurry. *Agricultural Wastes* 3(4): 311–321.

248. Wohlfarth, G. W. and Schroeder, G. L. 1979. Use of Manure in Fish Farming—A Review. *Agricultural Wastes* 1(4): 279–299.

READINGS IN HEALTH AND SAFETY

249. Amishima, M., et al. 1995. Dairy Farmers Have Increased Methacholine Bronchial Responsiveness Independent of Sensitization to Mold Antigens. *American Journal of Respiratory & Critical Care Medicine* 151: 1794–1798.

250. Bernues, A., et al. 1994. Farming Systems and Sanitary Problems in Mountain Cattle Farms. *Veterinary Research* 25(2–3): 305–312.

251. Chinabut, S., et al. 2006. Problems Associated with Shellfish Farming. *Reviews in Science & Technology* 25(2): 627–635.

252. Dalphin, J. C. H., et al. 1993. Etiologic Factors of Chronic Bronchitis in Dairy Farmers. *Chest* 103(2): 417–421.

253. Durborow, R. M. 1999. Health and Safety Concerns in Fisheries and Aquaculture. *Occupational Medicine* 14(2): 373–406.

254. Elliott, L., et al. 2004. Ecological Associations between Asthma Prevalence and Potential Exposure to Farming. *European Respiratory Journal* 24(6): 938–941.

255. Iversen, M., et al. 2000. Human Health Effects of Dust Exposure in Animal Confinement Buildings. *Journal of Agricultural Safety & Health* 6(4): 283–288.

256. Linden, A., et al. 2001. Cadmium in Organic and Conventional Pig Production. *Archives of Environmental Contamination & Toxicology* 40(3): 425–431.

257. McGinn, S. M., et al. 2003. Atmospheric Ammonia, Volatile Fatty Acids, and Other Odorants Near Beef Feedlots. *Journal of Environmental Quality* 32(4): 1173–1182.

258. Morrison H., et al. 1993. Farming and Prostate Cancer Mortality. *American Journal of Epidemiology* 137(3): 270–280.

259. Newnham, R. E. 1991. Agricultural Practices Affect Arthritis. *Nutrition & Health* 7: 89–100.

260. Oncul, O., et al. 2002. Cutaneous Anthrax in Turkey: A Review of 32 Cases. *Scandinavian Journal of Infectious Disease* 34(6): 413–416.

261. OSHA. 1993. *Ergonomics Program Management Guidelines for Meatpacking Plants* (OSHA 3123). Washington, DC: OSHA.

262. Perdue, M. L. and Swayne, D. E. 2005. Public Health Risk from Avian Influenza Viruses. *Avian Diseases* 49(3): 317–327.

263. Phillips, J. and French, M. 1999. State Regulation and the Hazards of Milk, 1900–1939. *Social History of Medicine* 12: 371–388.

264. Pratt, D. S. 1993. Respiratory Hazards in Agriculture: Beyond Dangerous Dust. *Seminars in Respiratory Medicine* 14(1): 8–14.

265. Spiewak, R. 2001. Pesticides as a Cause of Occupational Skin Diseases in Farmers. *Annals of Agricultural Environmental Medicine* 8: 1–5.

266. Vogelzang, P. F., et al. 1999. Organic Dust Toxic Syndrome in Swine Confinement Farming. *American Journal of Industrial Medicine* 35(4): 332–334.

267. Von Essen, S. G. and Auvermann, B. W. 2005. Health Effects from Breathing Air Near CAFOs for Feeder Cattle or Hogs. *Journal of Agromedicine* 10(4): 55–64.

268. Walsh, M. 2000. Farm Accidents: Their Causes and the Development of a Nurse-Led Accident Prevention Strategy. *Emergency Nurse* 8(7): 24–31.

269. Wells, S. J., et al. 1998. Key Health Issues for Dairy Cattle—New and Old. *Journal of Dairy Science* 81(11): 3029–3035.

113: Forestry and Logging
1131: Timber Tract Operations
1132: Forest Nurseries and Gathering of Forest Products
1133: Logging</ul>

READINGS IN HISTORY

270. Andrews, R. W. 1994. *Glory Days of Logging/Action in the Big Woods, British Columbia to California.* Atglen, PA: Schiffer Publishing.

271. Bacig, T. and Thompson, F. 1982. *Tall Timber: A Pictorial History of Logging in the Upper Midwest.* Bloomington, MN: Voyageur.

272. Becerra, C. A. 1994. Giants of the Swamp: The Story of South Florida's Logging Industry. *South Florida History Magazine* 22(1): 16–25.

273. Clark, A. H. 2002. Logging Life East of the Cascades: A Photo Essay. *Journal of the Shaw Historical Library* 16: 42–56.

274. Cowan, M. M. 2003. *Timberrr!: A History of Logging in New England.* Brookfield, CT: Millbrook Press.

275. Drushka, K. 1992. *Working in the Woods: History of Logging on the West Coast.* Maderia Park, BC: Harbour Publishing.

276. Drushka, K. and Konttinen, H. 1997. *Tracks in the Forest: The Evolution of Logging Machinery.* Maderia Park, BC: Harbour Publishing.

277. Fickle, J. E. 2004. Forest Products: The South's Forgotten Industry. *Journal of Mississippi History* 66(1): 1–16.

278. Herndon, G. M. 1979. Forest Products of Colonial Georgia. *Journal of Forest History* 23(3): 130–135.

279. Johnston, H. 1984. *The Whistles Blow No More: Railroad Logging in the Sierra Nevada, 1874–1942.* Glendale, CA: Trans-Anglo Books.

280. Karamanski, T. J. 1989. *Deep Woods Frontier: A History of Logging in Northern Michigan.* Detroit: Wayne State University Press.

281. LaLande, J. M. 1979. *Medford Corporation: A History of an Oregon Logging and Lumber Company.* Medford, OR: Medford Corp.

282. Mackenzie, D. E. and Maunder, E. R. 1972. Logging Equipment Development in the West. *Forest History* 16(3): 30–33.

283. Prouty, A. M. 1985. *More Deadly Than War! Pacific Coast Logging, 1827–1981*. New York: Garland.

284. Robbins, W. G. 1987. The Tarnished Dream: The Turbulent World of the Forest Products Industry in the Northwest. *Montana* 37(1): 63–65.

285. Rowley, W. D. and Rowley, P. 1992. Lumbering and Logging at Hobart Mills. *Nevada Historical Society Quarterly* 35(3): 195–201.

286. Smith, D. C. 1974. The Logging Frontier. *Journal of Forest History* 18(4): 96–106.

287. Uhler, R. S., ed. 1991. *Canada-United States Trade in Forest Products*. Vancouver: University of British Columbia Press.

READINGS IN HAZARDOUS WASTE MANAGEMENT AND POLLUTION PREVENTION

288. Beschta, R. L. *Wildfire and Salvage Logging: Recommendations for Ecologically Sound Post-Fire Salvage Logging and Other Post-Fire Treatments on Federal Lands in the West: A Report*. Eugene, OR: Pacific Rivers Council.

289. Boyer, J. N. and South, D. B. 1984. Forest Nursery Practices in the South. *Southern Journal of Applied Forestry*.

290. Brown, T. C., et al. 1993. Laws and Programs for Controlling Nonpoint Source Pollution in Forest Areas. *Water Resources Bulletin* 29(1): 1.

291. Diver, S. and Greer, L. 2000. *Sustainable Small-Scale Nursery Production*. Fayetteville, AR: National Center for Appropriate Technology.

292. Fisher, R. F., et al. 2000. *Ecology and Management of Forest Soils*. New York: John Wiley.

293. Hildebrand, D. M., et al. 2004. *Alternatives to Preplant Soil Fumigation for Western Forest Nurseries* (PNW-GTR-608). Washington, DC: USDA, Forest Service.

294. Ice, G. G., et al. 1997. 25 Years of the Clean Water Act: How Clean Are Forest Practices? *Journal of Forestry* 95(7): 9–13.

295. Juntunen, M-L., et al. 2002. Leaching of Nitrogen and Phosphorus during Production of Forest Seedlings in Containers. *Journal of Environmental Quality* 31: 1868–1874.

296. McKay, H. M. 1997. A Review of the Effect of Stresses between Lifting and Planting on Nursery Stock Quality and Performance. *New Forests* 13(1–3): 369–399.

297. South, D. B. 1996. Chemicals Used in Southern Forest Nurseries. *Southern Journal of Applied Forestry* 20(3): 127–135.

298. Wadleight, C. H. 1968. *Wastes in Relation to Agriculture and Forestry*. Washington, DC: Government Printing Office.

READINGS IN HEALTH AND SAFETY

299. Bordas, R. M., et al. 2001. Documentation of Hazards and Safety Perceptions for Mechanized Logging Operations in East Central Alabama. *Journal of Agricultural Safety & Health* 7(2): 113–123.

300. Egan, A. F. 1998. The Introduction of a Comprehensive Logging Safety Standard in the USA—The First Eighteen Months. *Journal of Forest Engineering* 9(1): 17–23.

301. ———. 1996. Hazards in the Logging Woods: Who's Responsible? *Journal of Forestry* 94(7): 16–20.

302. Helmkamp, J. C. and Derk, S. J. 1999. Nonfatal Logging-Related Injuries in West Virginia. *Journal of Occupational & Environmental Medicine* 41(11): 967–972.

303. Johnson, C. M., et al. 2002. When the Bough Breaks: A 10-Year Review of Logging Injuries Treated at a Rural Trauma Center in Pennsylvania. *American Surgeon* 68(7): 573–581.

304. Neitzel, R. and Yost, M. 2002. Task-based Assessment of Occupational Vibration and Noise Exposure in Forestry Workers. *AIHA Journal* 63(5): 617–627.

305. Nilsson, C. A., et al. 1987. Occupational Exposure to Chain Saw Exhausts in Logging Operations. *AIHA Journal* 48(2): 99–105.

306. NIOSH. 1995. *Preventing Injuries and Deaths of Loggers* (Pub. 95–101). Washington, DC: NIOSH.

307. Pratt, D. S. 1990. Occupational Health and the Rural Worker: Agriculture, Mining, and Logging. *Journal of Rural Health* 6(4): 399–417.

308. Scott, D. F. 2004. A Study of Logger Fatalities from 1992–2000. *Injury Prevention* 10(4): 239–243.

309. Smith, S., et al. 1999. Logging Injuries in Louisiana: Nature, Trends, and Rehabilitation Considerations. *Work* 12(3): 261–273.

310. Struttmann, T. W. and Scheerer, A. L. 2001. Fatal Injuries Caused by Logs Rolling Off Trucks. *American Journal of Industrial Medicine* 39(2): 203–208.

311. Wasterlund, D. S. 1998. A Review of Heat Stress Research with Application to Forestry. *Applied Ergonomics* 29(3): 179–183.</bib>

114: Fishing, Hunting, and Trapping

READINGS IN HISTORY

312. Allen, W. E. 1983. Eagle Trapping along the Little Missouri River. *North Dakota History* 50(1): 4–22.

313. Bailey, B. C. 2003. Brooke County Trapping. *Goldenseal* 29(4): 49–51.

314. Bogue, M. B. 2000. *Fishing the Great Lakes: An Environmental History*. Madison: University of Wisconsin Press.

315. Donlon, J. G. 1996. Hunting for Leisure: The Social Components and Historic Foundations of American Sport Hunting. Dissertation, University of Illinois, Urbana-Champaign.

316. Dunlap, T. R. 1988. Sport Hunting and Conservation, 1880–1920. *Environmental Review* 12(1): 51–60.

317. Flanagan, J. T. 1979. Hunting in Early Illinois. *Journal of the Illinois State Historical Society* 72(1): 2–12.

318. Gohdes, C., ed. 1965. Hunting in the Old South. *Georgia Review* 19(3): 350–359.

319. Heatwole, C. and West, N. 1985. Shorefront Fishing in New York City. *Geographical Review* 75(3): 245–264.

320. Herman, D. J. 2001. *Hunting and the American Imagination*. Washington, DC: Smithsonian Institution Press.

321. Hogan, A. S. 1985. An Introduction to the History of Fly Fishing in America. *American Fly Fisher* 12(4): 2–9.

322. Holden, R. J. and Holden, D. J. 2000. *The Hunting Pioneers, 1720–1840: Ultimate Backwoodsmen on the Early American Frontier*. Bowie, MD: Heritage Books.

323. Jones, C. 1973. Fox Hunting in America. *American Heritage* 24(6): 62–68.

324. Kaups, M. 1978. North Shore Commercial Fishing, 1849–1870. *Minnesota History* 46(2): 43–58.

325. Leuth, F. X. 1968. The Decline in Trapping in Alabama. *Journal of the Alabama Academy of Science* 39(1): 42–47.

326. Lund, J. 1995. *Flatheads and Spooneys: Fishing for a Living in the Ohio River Valley.* Lexington: University Press of Kentucky.

327. Mighetto, L. 1995. Sport Fishing on the *Columbia* River. *Pacific Northwest Quarterly* 87(1): 5–15.

328. Muldoon, D. D. 1987. Daily Life of the Mountain Trapper. *Journal of the West* 26(4): 14–20.

329. Noon, J. 2003. *Fishing in New Hampshire: A History*. Warner, NH: Moose Country.

330. Petersen E. T. 1979. *Hunter's Heritage: A History of Hunting in Michigan*. Lansing: Michigan United Conservation Clubs.

331. Philips, R. H. 1974. The Present and Future of the West Coast Fishing Industries. *U.S. Naval Institute Proceedings* 100(5): 190–203.

332. Schug, D. M. 2001. Hawaii's Commercial Fishing Industry: 1820–1945. *Hawaiian Journal of History* 35: 15–34.

333. Schullery, P. 1987. *American Fly Fishing: A History.* New York: Nick Lyons Books.

334. ———. The Colonial Angler. *American Fly Fisher* 14(2): 2–5.

335. Vernon, S. K. 1990. Fishing around Philadelphia. *Pennsylvania Heritage* 16(2): 24–31.

336. Underhill, L. E. and Littlefield, D. F. 1971. Quail Hunting: Big Business in Early Oklahoma. *Chronicles of Oklahoma* 49(3): 315–333.

337. White, L. H. and Gowans, F. R. 1993. Traders to Trappers: Andrew Henry and the Rocky Mountain Fur Trade. *Montana* 43(3): 54–63.

READINGS IN HAZARDOUS WASTE MANAGEMENT AND POLLUTION PREVENTION

338. Georgia Department of Natural Resources. 2000. *Dove Hunting and Agricultural Practices in Georgia*. Athens, GA: Digital Library of Georgia

339. Moore G., and Jennings S. 2000. *Commercial Fishing: The Wider Ecological Impacts*. London: Blackwell.

340. Pohland, F. G. and Hudson, J. W. 1978. *Wastewater Management Alternatives for the Shellfish Processing Industry*. Savannah, GA: Georgia Marine Science Center, University System of Georgia.

341. Thomas, V. G. and Guitart, R. 2003. Lead Pollution from Shooting and Angling, and a Common Regulative Approach. *Environmental Policy & Law* 33(3/4): 143–149.

READINGS IN HEALTH AND SAFETY

342. Anda, P., et al. 2001. Waterborne Outbreak of Tularemia Associated with Crayfish Fishing. *Emerging Infectious Disease* 7(3 Supp.): 575–582.

343. Burke W. A., et al. 2006. Skin Problems Associated to the Occupation of Commercial Fishing in North Carolina. *North Carolina Medical Journal* 67(4): 260–265.

344. Carter, G. L. 1989. Accidental Firearm Fatalities and Injuries among Recreational Hunters. *Annals of Emergency Medicine* 18(4): 406–409.

345. Chalmers, D. and Morrison, L. 2003. Epidemiology of Non-submersion Injuries in Aquatic Sporting and Recreational Activities. *Sports Medicine.* 33(10): 745–770.

346. Chi, L., et al. 1996. Electrical Burns from Graphite Fishing Rods. *Burns* 22(8): 638–640.

347. Dieter, R. A., et al. 2001. Bear Mauling: A Descriptive Review. *International Journal of Circumpolar Health* 60(4): 696–704.

348. Dudkiewicz, I., et al. 2000. Fishing Penetration Injuries. *British Journal of Sports Medicine* 34(6): 459–461.

349. Gammons, M. G. and Jackson, E. 2001. Fishhook Removal. *American Family Physician* 63(11): 2231–2236.

350. Hastein, T., et al. 2006. Food Safety Hazards That Occur during the Production Stage: Challenges for Fish Farming and the Fishing Industry. *Reviews in Science & Technology* 25(2): 607–625.

351. Hollerman, J. J. 1988. Gunshot Wounds. *American Family Physician* 37(5): 231–246.

352. Marshall, S. W., et al. 2004. Work-related Injuries in Small-scale Commercial Fishing. *Injury Prevention* 10(4): 217–221.

353. Matheson, C., et al. 2001. The Health of Fisherman in the Catching Sector of the Fishing Industry: A Gap Analysis. *Occupational Medicine* 51(5): 305–311.

354. Minciullo, P. L., et al. 2004. Contact Dermatitis from a Fishing Rod. *Contact Dermatitis* 50(5): 322.

115: Support Activities for Agriculture and Forestry
1151: Support Activities for Crop Production
1152: Support Activities for Animal Production
1153: Support Activities for Forestry

READINGS IN HISTORY

355. Adams, W. H. 1990. Landscape Archaeology, Landscape History, and the American Farmstead. *Historical Archaeology* 24(4): 92–101.

356. Aiken, C. S. 1973. The Evolution of Cotton Ginning in the Southeastern United States. *Geographical Review* 63(2): 196–224.

357. Britton, K. G. 1991. Bale O' Cotton: Cotton Ginning and Oral Tradition. *Purview Southwest* 61–66.

358. ——. 1992. *Bale O' Cotton: The Mechanical Art of Cotton Ginning*. College Station: Texas A&M University Press.

359. Conlogue, W. 2001. Managing the Farm, Educating the Farmer: O Pioneers! And the New Agriculture. *Great Plains Quarterly* 21(1): 3–15.

360. Drache, H. M. 1985. *Plowshares to Printouts: Farm Management as Viewed through 75 Years of the Northwest Farm Managers Association*. Danville, IL: Interstate.

361. Heikkonen, E. 1995. *Reaping the Bounty: McCormick Harvesting Machine Company Turns Abroad, 1878–1902*. Helsinki: Finnish Historical Society.

362. Hildreth, R. J. and Enstes, D. P. 2000. Farm Foundation's Role as a Policy Issues Catalyst. *Agricultural History* 74(2): 352–365.

363. Hovland, M. 1994. The Cotton Ginning Reports Program at the Bureau of the Census. *Agricultural History* 68(2): 146–156.

364. Lakwete, A. 1997. Cotton Ginning in America: 1780–1860. Dissertation, University of Delaware.

365. Queen, G. S. 1964. The McCormick Harvesting Machine Company in Russia. *Russian Review* 23(2): 164–181.

366. Sawer, B. J. 1973. Predictors of the Farm Wife's Involvement in General Management and Adoption Decisions. *Rural Sociology* 38(4): 412–426.

367. Vandergriff, A. L. 1997. *Ginning Cotton: An Entrepreneur's Story.* Lubbock: Texas Tech University Press.

368. White, R. E. 1957. Cotton Ginning in Texas to 1861. *Southwestern Historical Quarterly* 61(2): 257–269.

369. ——. 1967. The Texas Cotton Ginning Industry, 1860–1900. *Texana* 5(4): 344–358.

370. Wrenn, L. B. 1994. Cotton Gins and Cottonseed Oil Mills in the New South. *Agricultural History* 68(2): 232–242.

READINGS IN HAZARDOUS WASTE MANAGEMENT AND POLLUTION PREVENTION

371. Anthony, W. S. and Mayfield, W. D. 1994. *Cotton Ginners Handbook.* Washington, DC: Agricultural Research Service.

372. Potter, W. D., et al. 2000. Aerial Spray Deposition Management Using the Genetic Algorithm. *Lecture Notes in Computer Science* 1821: 210–219.

373. Sappington, T. W., et al. 2004. Boll Weevil Survival through Cotton Gin Trash Fans. *Journal of Economic Entomology* 97(5): 1612–1618.

374. Singh, G. and Iyer, G. V. 2004. Environmental Impact Studies of Chrome Rollers Used by Cotton Roller Ginning Industries and Design and Development of Pollution Free Chromeless RCF Rollers. *Environmental Monitoring & Assessment* 96(1–3): 163–181.

READINGS IN HEALTH AND SAFETY

375. Anthony, W. S. and Columbus, E. P. 1985. Repairable Dust and Gin Machinery. *Applied Ergonomics* 16(3): 201–207.

376. Bethea, R. M. and Morey, P. R. 1977. Occupational Dust Exposures in West Texas Cotton Gins. *AIHA Journal* 38(6): 284–294.

377. Bethea, R. M., et al. 1978. Evaluation of a Fluorescent Dust Tracer Technique in Cotton Ginning. *AIHA Journal* 39(12): 998–1008.

378. Brown, J. N., et al. 2000. Herbicide Spray Drift Odor: Measurement and Toxicological Significance. *Archives of Environmental Contamination & Toxicology* 38(3): 390–397.

379. North, J. P. and Hightower, R. D. 1957. Crop Dusting Injuries. *American Journal of Surgery* 93(4): 609–612.

380. Palmer, A., et al. 1978. Byssinosis and Chronic Respiratory Disease in U.S. Cotton Gins. *Journal of Occupational Medicine* 20(2): 96–102.

381. Rafnsson, V. 2006. Cancer Incidence among Farmers Exposed to Lindane while Sheep Dipping. *Scandinavian Journal of Work & Environmental Health* 32(3): 185–189.

382. Seiber, J. N., et al. 1979. Residues of Toxaphene, DEF, and Paraquat in Plant Parts and Gin Waste from a Treated Cotton Field. *Archives of Environmental Contamination & Toxicology* 8(2): 125–137.

383. Stephens, R., et al. 1995. Neuropsychological Effects of Long-term Exposure to Organophosphates in Sheep Dip. *Lancet* 345(8958): 1135–1139.

Sector 21: Mining

211: Oil and Gas Extraction

READINGS IN HISTORY

1. Adelman, M. A. 1985. Western Hemisphere Perspectives: Oil and Natural Gas. *Contemporary Policy Issues* 3(4): 2–12.
2. Barger, H. and Schurr, S. H. 1975. *The Mining Industries, 1899–1939: A Study of Output, Employment, and Productivity*. Manchester, NH: Ayer Publishing.
3. Bayley, W. S. 1910. *Iron Mines and Mining in New Jersey*. MacCreillish & Quigley, State Printers.
4. Black, B. 1997. Recasting the Unalterable Order of Nature: Photography and the First Oil Boom. *Pennsylvania History* 64(2): 275–299.
5. Blomstedt, L. W. 2002. Frank "Doc" Dotterweich: The Ambassador of Natural Gas Engineering. *Journal of South Texas* 15(1): 73–97.
6. Branch M.C. 1972. Oil Extraction, Urban Environment, and City Planning. *Journal of the American Institute of Planners* 38(3): 140–154.
7. Castaneda, C. J. 1999. *Invisible Fuel: Manufactured and Natural Gas in America, 1800–2000*. New York: Twayne.
8. ——. 2000. *Invisible Fuel: Manufactured and Natural Gas in America, 1800–2000*. New York: Twayne.

9. ——. 2004. History beneath the surface: Natural Gas Pipelines and the National Historic Preservation Act. *Public Historian* 26(1): 105–121.

10. Clark, B. 1977. The Beginning of Oil and Gas Conservation in Oklahoma, 1907–1931. *Chronicles of Oklahoma* 55(4): 375–391.

11. Dean, C. J. 1974. Gas Developments in the United States. *Geography* 59(1): 41–45.

12. Foss, M. M. 1995. Natural Gas in the 21st Century. Dissertation, University of Houston.

13. Frank, H. J. and Schanz, J. J. 1973. The Future of American Oil and Natural Gas. *Annals of the American Academy of Political & Social Science* 410: 24–34.

14. Frickle, S. and Freudenberg, W. R. 1996. Mining the Past: Historical Context and the Changing Implications of Natural Resource Extraction. *Social Problems* 43(4): 444–466.

15. Friedman, B. L. 1992. Productivity in Crude Oil and Gas Production. *Monthly Labor Review* 115(3): 9–17.

16. Glass, J. A. 2000. The Gas Boom in East Central Indiana. *Indiana Magazine of History* 96(4): 313–335.

17. Greene, R. 1981. Employment Trends in Energy Extraction. *Monthly Labor Review* 104(5): 3–8.

18. Hair, W. I. 1977. Plundered Legacy: The Early History of North Louisiana's Oil and Gas Industry. *North Louisiana Historical Association Journal* 8(5): 179–184.

19. Hatheway, A. W. 2002. Geoenvironmental Protocol for Site and Waste Characterization of Former Manufactured Gas Plants; Worldwide Remediation Challenge in Semi-volatile Organic Wastes. *Engineering Geology* 64: 317–338.

20. Libecap, G. D. 1989. The Political Economy of Crude Oil Cartelization in the United States, 1933–1972. *The Journal of Economic History* 49(4): 833–855.

21. Libecap, G. D. and Wiggins, S. N. 1984. Contractual Responses to the Common Pool: Prorationing of Crude Oil Production. *American Economic Review* 74(1): 87–98.

22. Prindle, D. F. 1981. The Texas Railroad Commission and the Elimination of the Flaring of Natural Gas, 1930–1949. *Southwestern Historical Quarterly* 84(3): 293–308.

23. Reis, M. C. 2001. By the Gasworks Wall: The Split Image of the Pioneer American Gas Industry and Its Buildings. *Pioneer America Society Transactions* 24: 83–95.

24. Sanders, E. M. 1981. *The Regulation of Natural Gas: Policy and Politics, 1938–1978*. Philadelphia: Temple University Press.

25. Screws, D. E. 2001. A History of Natural Gas Processing. *Permian Historical Annual* 41: 28–43.

26. Troesken, W. 1993. The Chicago Gas Industry, 1887–1913: Economic Analysis of Institutional Change. Dissertation, Washington University.

27. Weaver, B. D. 1989. The Discovery and Development of the Panhandle Oil and Gas Field. *Journal of the West* 28(4): 4–9.

28. Wickstrom, L. H. and Gray, J. D. 1994. Boom Towns: Oil and Gas in Northwestern Ohio. *Timeline* 11(6): 2–15.

29. Williams, J. C. 1996. Fuel at Last: Oil and Gas for California, 1860s–1940s. *California History* 75(2): 114–127.

READINGS IN HAZARDOUS WASTE MANAGEMENT AND POLLUTION PREVENTION

30. Gray, J. S. 2002. Perceived and Real Risks: Produced Water from Oil Extraction. *Marine Pollution Bulletin* 44(11): 1171–1172.

31. O'Rourke, D. and Connolly, S. 2003. Just Oil? The Distribution of Environmental and Social Impacts of Oil Production and Consumption. *Annual Review of Energy and the Environment* 28: 587–618.

32. Reis, J. C. 1996. *Environmental Control in Petroleum Engineering*. Houston, TX: Gulf Publishing.

33. Sittig, M. 1978. *Petroleum Transportation and Production: Oil Spill and Pollution Control*. Park Ridge, NJ: Noyes Data Corporation.

34. Texas Railroad Commission. 1999. *Waste Minimization in the Oil Field*.

35. U.S. Environmental Protection Agency. 2000. *Profile of the Oil and Gas Extraction Industry* (EPA/310/R-00/004). Washington, DC: Office of Enforcement & Compliance Assurance.

READINGS IN HEALTH AND SAFETY

36. Buffler, P. A., et al. 2004. Primary Brain Tumor Mortality at a Petroleum Exploration and Extraction Research Facility. *Journal of Occupational & Environmental Medicine* 46(3): 257–270.

37. Duffy, B. 1996. Dental Problems in the Offshore Oil and Gas Industry: A Review. *Occupational Medicine* 46(1): 79–83.

38. Mackerer, C. R., et al. 2003. Petroleum Mineral Oil Refining and Evaluation of Cancer Hazard. *Applied Occupational & Environmental Hygiene* 18(11): 890–901.

39. Moen, B. E. 2004. Occupational Environment and Health in Oil Industry. *Tidsskr. Nor. Laegeforen* 124(20): 2599

40. Morken, T., et al. 2004. Musculoskeletal Disorders in the Offshore Oil Industry. *Tidsskr. Nor. Laegeforen* 124(20): 2623–2626.

212: Mining (Except Oil and Gas)
2121: Coal Mining
2122: Metal Ore Mining
2123: Nonmetallic Mineral Mining and Quarrying

READINGS IN HISTORY

41. Aldrich, M. 1997. The Perils of Mining Anthracite: Regulation, Technology and Safety 1870–1945. *Pennsylvania History* 64(3): 361–383.

42. Amundson, M.A. 1995. Home on the Range No More: The Boom and Bust of a Wyoming Uranium Mining Town, 1957–1988. *Western Historical Quarterly* 26(4): 483–505.

43. ——. 2002. *Yellowcake Towns: Uranium Mining Communities in the American West.* Boulder, CO: University Press of Colorado.

44. Anonymous. 1974. Lignite, Strip Mining, and Reclamation: A Bibliography. *North Dakota History* 41(3): 14–19.

45. Aurand, H. W. 1980. The Anthracite Miner: An Occupational Analysis. *Pennsylvania Magazine of History and Biography* 104(4): 462–473.

46. ——. 1985. Mine Safety and Social Control in the Anthracite Industry. *Pennsylvania History* 52(4): 227–241.

47. Badash, L. 1978. Radium, Radioactivity, and the Popularity of Scientific Discovery. *Proceedings of the American Philosophical Society* 122(3): 145–154.

48. Balsley, H. W. 1991. Early Days of Uranium. *Canyon Legacy* 10: 2–7.

49. Barnett, L. 2002. Haulin' Ash: 100 Years of Making Potash. *Michigan History Magazine* 86(1): 16–21.

50. Beauregard, E. E. 1998. L. Milton Ronsheim and Strip Coal Mining in Ohio. *Journal of Unconventional History* 9(3): 16–33.

51. Blakey, A. F. 1973. The Florida Phosphate Industry: A History of the Development and Use of a Vital Mineral. Cambridge, MA: Harvard University Press.

52. Burke, J. A. 1975. Barrels to Barrows, Buckets to Belts: 120 Years of Iron Ore Handling on the Great Lakes. *Inland Seas* 31(4): 266–277.

53. Carlson, A. W. 1972. Lignite Coal as an Enabling Factor in the Settlement of Western North Dakota. *Great Plains Journal* 11(2): 145–153.

54. Chadwick, R. A. 1982. Montana's Silver Mining Era: Great Boom and Great Bust. *Montana* 32(2): 16–31.

55. Chandler, A. D. 1972. Anthracite Coal and the Beginnings of the Industrial Revolution in the United States. *Business History Review* 46(2): 141–181.

56. Chandonnet, A. 1979. Evan Jones and Alaska Coal Mining. *Alaska Journal* 9(2): 92–94.

57. Corbett, W. P. 1978. The Red Pipestone Quarry. *South Dakota History* 8(2): 99–116.

58. DeSanto, J. 1985. Drilling at Kintla Lake: Montana's First Oil Well. *Montana* 35(1): 24–37.

59. DiCiccio, C. P. 1998. The Rise and Fall of King Coal: A History of the Bituminous Coal and Coke Industry of Pennsylvania from 1740–1945. Dissertation, University of Pennsylvania.

60. Dix, K. 1988. *What's a Coal Miner to Do? The Mechanization of Coal Mining.* Pittsburgh: University of Pittsburgh Press.

61. Eckman, D. J. 1998. Early Silver Mining in Lancaster County. *Journal of the Lancaster County Historical Society* 100(4): 405–410.

62. Fernandes, R. C. and Carvalho, F. M. 2000. Intervertebral Disk Disease among Oil Drilling Workers. *Cad Saude Publica* 16(3): 661–669.

63. Fishback, P. V. 1986. Workplace Safety during the Progressive Era: Fatal Accidents in Bituminous Coal Mining, 1912–1923. *Explorations in Economic History* 23(3): 269–298.

64. Fountain, D. 1992. *Michigan Gold Mining in the Upper Peninsula.* Duluth: Lake Superior Port Cities.

65. Gardner, D. A. and Flores, V. R. 1989. *Forgotten Frontier: A History of Wyoming Coal Mining*. Boulder, CO: Westview Press.

66. Gorman, M. 1986. Chronicle of a Silver Mine: The Meadow Valley Mining Company of Pioche. *Nevada Historical Society Quarterly* 29(2): 69–88.

67. Graebner, W. 1974. Great Expectations: The Search for Order in Bituminous Coal 1890–1917. *Business History Review* 48(1): 49–72.

68. Grieve, A. 1988. Oil: Toxicity of Drilling Mud. *Occupational Health* 40(12): 736–739.

69. Grindle, R. L. 1977. *Tombstones and Paving Blocks: The History of the Maine Granite Industry*. Rockland: Courier of Maine.

70. Hakala, D. R. 1969. The Iron Ore Industry: A Study in Shifts in Ownership and Control. *Quarterly Review of Economics & Business* 9(1): 45–52.

71. Hawley, E. W. 1968. Secretary Hoover and the Bituminous Coal Problem, 1921–1928. *Business History Review* 429(3): 247–270.

72. Henderson, D. F. 1964. The Texas Coal Mining Industry. *Southwestern Historical Quarterly* 68(2): 207–219.

73. Herendorf, C. E. 2004. History of the Lake Erie Sand and Gravel Industry. *Inland Seas* 60(3): 188–205.

74. Hightower, M. J. 1984. Cattle, Coal and Indian Land: A Tradition of Mining in Southeastern Oklahoma. *Chronicles of Oklahoma* 62(1): 4–25.

75. Hines, E. and Smith, M. S. 2002. Gold Is Where You Find It: Placer Mining in North Carolina, 1799–1849. *Earth Sciences History* 21(2): 119–149.

76. Hoffman, J. N. 1978. Pennsylvania's Bituminous Coal Industry: An Industry Review. *Pennsylvania History* 45(4): 351–363.

77. Horton, J. 1977. Phosphate. *Idaho Heritage* 7: 44–47.

78. Jensen, R. E. 1987. Nebraska's World War I Potash Industry. *Nebraska History* 68(1): 28–42.

79. Johnson, J. P. 1979. *The Politics of Soft Coal: The Bituminous Industry from World War I through the New Deal*. Urbana: University of Illinois Press.

80. Johnson, K. R. 1970. Gold Mining in Alabama in the 20th Century. *Journal of the Alabama Academy of Science* 41(1): 42–48.

81. Jones, W. R. 1984. Coal Mine Explosions at Almy, Wyoming: Their Influence on Wyoming's First Coal Mining Safety Laws. *Annal of Wyoming* 56(1): 54–65.

82. Kassner, J. 1996. The Mining of Port Jefferson Harbor for Sand and Gravel, 1910–1948. *Long Island Historical Journal* 8(2): 213–224.

83. Kelley, K. A. and Davies, J. K. 1983. Minersville: The Beginnings of Lead-silver Mining in Utah. *Utah Historical Quarterly* 51(3): 229–245.

84. Knapp, R. F. and Glass, B. D. 1999. *Gold Mining in North Carolina: A Bicentennial History*. Raleigh: North Carolina Division of Archeology & History.

85. Koplos, J. J. 1992. An Econometric Analysis of Mine Production Fluctuations: A Study of Historic Wisconsin Zinc Mining. Dissertation, University of Wisconsin, Madison.

86. Krueger, J. R. 1967. Indiana Limestone Industry Terms. *American Speech* 42(4): 289–296.

87. LeCain, T. J. 1998. Moving Mountains: Technology and the Environment in Western Copper Mining. Dissertation, University of Delaware.

88. Long, J. D. 1976. The Nickel Mines of Lancaster County. *Journal of the Lancaster County Historical Society* 80(3): 157–177.

89. Longin, T. C. 1974. Coal, Congress, and the Courts: The Bituminous Coal Industry and the New Deal. *West Virginia History* 35(2): 101–130.

90. Maddala, G. S. 1965. Productivity and Technological Change in the Bituminous Coal Industry, 1919–1954. *Journal of Political Economy* 73(4): 352–365.

91. Malley, M. 2002. Bygone Spas: The Rise and Decay of Oklahoma's Radium Water. *Chronicles of Oklahoma* 80(4): 446–467.

92. Margolis, E. 1985. Western Coal Mining as a Way of Life: An Oral History of the Colorado Coal Miners to 1914. *Journal of the West* 24(3): I-115.

93. McKinley, S. W. 2004. *T*he Origins of "King Phosphate" in the New South: Workers, Managers, and Entrepreneurs in South Carolina's Phosphate and Fertilizer Industries, 1865–1884. Dissertation, University of Delaware.

94. Merk, G. P. 1996. The Genesis of Michigan's Iron Ore. *Michigan Academician* 28(2): 165–180.

95. Miller, H. 1980. Potash from Wood Ashes: Frontier Technology in Canada and the United States. *Technology and Culture* 21(2): 187–208.

96. Mogren, E. W. 1995. Warm Sand: Public Policy and the Uranium Milling Industry 1948–1978. Dissertation, University of Michigan.

97. Myers, R. C. 2000. Railroads, Stone Quarries and the Colorado State Capitol. *Journal of the West* 39(2): 37–45.

98. ——. 2002. No One Stayed: Quarrying Granite in Aberdeen, Colorado, 1889–1892. *Mining History* 9: 54–64.

99. Naske, C. M. 1996. Mining Coal on the Meade River, Alaska. *Pacific Northwest Quarterly* 88(1): 3–12.

100. Navin, T. R. 1978. *Copper Mining and Management*. Tucson: University of Arizona Press.

101. Norgaard, R. B. and Gwo, J. L. 1986. Petroleum Accessibility and Drilling Technology: An Analysis of U.S. Development Costs from 1959 to 1978. *Land Economics* 62(1): 14–25.

102. Oihus, C. A. 1978. Lignite: North Dakota's Fledgling Coal Industry, 1873–1900. *North Dakota Quarterly* 46(4): 51–67.

103. Oremerod, A. D., et al. 1998. Novel Causes of Contact Dermatitis from Offshore Oil-based Muds. *Contact Dermatitis* 39(5): 262–263.

104. Papa, C. 2002. The New Jersey Zinc Company: Womb to Tomb in Two Mining Towns. *New Jersey History* 120(1–2): 58–90.

105. Pearre, N. C. 1964. Mining for Copper and Related Minerals in Maryland. *Maryland Historical Magazine* 59(1): 15–33.

106. Peterson, J. M. 1979. The Logan County Nickel Mine. *Kansas History* 2(1): 26–33.

107. Pitcaithley, D. 1978. Zinc and Lead Mining Along the Buffalo River. *Arkansas Historical Quarterly* 37(4): 293–305.

108. Powell, H. B. 1980. The Pennsylvania Anthracite Industry, 1769–1976. *Pennsylvania History* 47(1): 3–28.

109. Rawls, J. J. and Orsi, R. J., eds. 1999. *A Golden State: Mining and Economic Development in Gold Rush California.* Berkeley: University of California Press.

110. Ringholz, R. C. 1989. *Uranium Frenzy: Boom and Bust on the Colorado Plateau*. New York: Norton.

111. Rintoul, W. 1981. *Drilling Ahead: Tapping California's Richest Oil Fields*. Santa Cruz: Western Tanager.

112. Roark, M. 1982. Early American Lead Mining in Southeast Missouri. *Pioneer America Society Transactions* 5: 55–62.

113. ——. 1988. The Effect of Lead and Iron Mining on the Cultural Development of Eastern Missouri, 1860–1880. *Pioneer America Society Transactions* 11: 9–17.

114. Roppel, P. 1979. View Cove, a Limestone Quarry. *Alaska Journal* 9(2): 44–48.

115. Rosenquist, V. B. 1987. The Iron Ore Eaters: A Portrait of the Mining Community of Moriah, New York. Dissertation, Duke University.

116. Rossel, J. 2002. Industrial Structure, Union Strategy, and Strike Activity in American Bituminous Coal Mining, 1881–1894. *Social Science History* 26(1): 1–32.

117. Schneider, D. 1983. *Black Diamonds: Life and Work in Iowa's Coal Mining Communities, 1895–1925*. Ames: Iowa State University Press.

118. Sewell, S. 1992. Amongst the Damp: The Dangerous Profession of Coal Mining in Oklahoma, 1870–1935. *Chronicles of Oklahoma* 70(1): 66–83.

119. Sorensen, D. 1963. Wonder Mineral: Utah's Uranium. *Utah Historical Quarterly* 31(3): 280–290.

120. Spude, R. L. 1974. A Land of Sunshine and Silver: Silver Mining in Central Arizona, 1871–1885. *Journal of Arizona History* 16(1): 29–76.

121. Szwilski, A. B. 1987. An Analysis of U.S. Coal Industry Trends, 1900–1986. *Mining Science & Technology* 6(1): 77–87.

122. Wallace, M. 1987. Dying for Coal: The Struggle for Health and Safety Conditions in American Coal Mining, 1930–1982. *Social Forces* 66(2): 336–364.

123. Whiteside, J. 1989. Coal Mining, Safety, and Regulation in New Mexico, 1882–1933. *New Mexico Historical Quarterly* 64(2): 159–184.

124. Williams, D. 1992. Dawn of a New Era: Gold Mining in 20th Century Georgia. *Georgia Historical Quarterly* 76(4): 915–928.

125. Wines, R. A. 1985. *Fertilizer in America: From Waste Recycling to Resource Exploitation.* Philadelphia: Temple University Press.

126. Wood, T. S. 1975. A Brief History of the Granite Industry in San Diego County. *Journal of San Diego History* 20(3): 61–66.

127. Wright, W. A. 1979. *The Granite Town: Milford, New Hampshire, 1901–1978*. Phoenix: Canaan.

128. Zoeliner, T. 2000. The Uranium Rush. *American Heritage of Invention & Technology* 16(1): 56–63.

READINGS IN HAZARDOUS WASTE MANAGEMENT, AND POLLUTION PREVENTION

129. Akcil, A. and Koldas, S. 2006. Acid Mine Drainage: Causes, Treatment, and Case Studies. *Journal of Cleaner Production* 14 (12–13): 1139–1145.

130. Chang, C. T. 2004. Assessment of Influential Range and Characteristics of Fugitive Dust in Limestone Extraction Processes. *JAWMA* 54(2): 141–148.

131. El-Fadel, M., et al. 2001. Environmental Management of Quarries as Waste Disposal Facilities. *Environmental Management* 27(4): 515–531.

132. Fields, S. 2001. Tarnishing the Earth: Gold Mining's Dirty Secret. *Environmental Health Perspectives*. 109(10): 474–481.

133. Kihlstedt, P. G. 1972. Waste Water in Metal Mining Industries. *Pure Applied Chemistry* 29(1): 323–332.

134. Lottermoser, B. G. 2003. *Mine Wastes: Characterization, Treatment and Environmental Impacts.* New York: Springer.

135. Low, N. and Gleeson, B. 1998. Situating Justice in the Environment: The Case of BHP at the OK Tedi Copper Mine. *Antipode* 30(3): 201–226.

136. McElfish, J. M. 1996. *Hard Rock Mining: State Approaches to Environmental Protection*. Washington, DC: Environmental Law Institute.

137. McElfish, J. M. and Beier, A. E. 1990. *Environmental Regulation of Coal Mining: SMCRA's Second Decade*. Washington, DC: Environmental Law Institute.

138. Meador, M. R. and Layher, A. O. 1998. Instream Sand and Gravel Mining: Environmental Issues and Regulatory Process in the United States. *Fisheries* 23(11): 6–13.

139. Muezzinoglu, A. 2003. A Review of Environmental Considerations on Gold Mining and Production. *Critical Reviews in Environmental Science & Technology* 33(1): 45–72.

140. Porter, K. E. and Bleiwas, D. I. 2003. *Physical Aspects of Waste Storage from a Hypothetical Open Pit Porphyry Copper Operation* (OFR 03–143). U.S. Geological Survey. http://pubs.usgs.gov/of/2003/of03-143 (accessed 15 October 2008).

141. Purcell, T. W. and Peters, J. J. 1999. Historical Impacts of the Environmental Regulation of Silver. *Environmental Toxicology & Chemistry* 18(1): 3–8.

142. Rosner, U. 1998. Effects of Historical Mining Activities on Surface Water and Groundwater—An Example from Northwest Arizona. *Environmental Geology* 33(4): 224–230.

143. Sengupta, M. 1993. *Environmental Impacts of Mining: Monitoring, Restoration, and Control.* Boca Raton, FL: Lewis Publishers.

144. U.S. Environmental Protection Agency. 1995. *Profile of the Stone, Clay, Glass and Concrete Products Industry* (EPA/310/R-95/017). Washington, DC: EPA.

145. Warhurst, A. 1999. *Mining and The Environment: Case Studies from the Americas.* Ottawa, ON: International Development Research Centre.

146. Younger, P. L., et al. 2002. *Mine Water: Hydrology, Pollution, Remediation*. Boston: Kluwer Academic Publishers.

READINGS IN HEALTH AND SAFETY

147. Amandus, H. E., et al. 1989. Health Status of Anthracite Surface Coal Miners. *Archives of Environmental Health* 44(2): 75–81.

148. Argust, P. 1998. Distribution of Boron in the Environment. *Biological Trace Element Research* 66(1–3): 131–143.

149. Aslibekina, O. and Moles, R. 2003. Environmental Risk Assessment of Metals-Contaminated Soils at Silver Mines Abandoned Mine Site. *Environmental Geochemistry & Health* 25(2): 247–266.

150. Cashdollar, K. L. 1996. Coal Dust Explosibility. *Journal of Loss Prevention in Processing Industries* 9(1): 65–76.

151. Clark, C. 1993. Radium Poisoning Revealed. *Humboldt Journal of Social Relations* 19(1): 73–116.

152. Coggon, D. and Taylor, A. N. 1998. Coal Mining and Chronic Obstructive Pulmonary Disease: A Review of the Evidence. *Thorax* 53: 398–407.

153. Davidson, R. G., et al. 1988. Occupational Hygiene Aspects of the Use of Oil-based Drilling Fluids. *Annals of Occupational Hygiene* 32(3): 325–332.

154. Donoghue, A. M. 2004. Occupational Health Hazards in Mining: An Overview. *Occupational Medicine* 54: 283–289.

155. Eide, I. 1990. A Review of Exposure Conditions and Possible Health Effects Associated with Aerosol and Vapor from Low Aromatic Oil-Based Drilling Fluids. *Annals of Occupational Hygiene* 34(2): 149–157.

156. Eisler, R. 2003. Health Risks of Gold Miners: A Synoptic Review. *Environmental Geochemistry & Health* 25(3): 325–345.

157. ———. 2004a. Arsenic Hazards from Gold Mining to Humans, Plants, and Animals. *Reviews in Environmental Contamination & Toxicology* 180: 133–165.

158. ———. 2004b. Mercury Hazards from Gold Mining to Humans, Plants, and Animals. *Reviews in Environmental Contamination & Toxicology* 181: 139–198.

159. Field, G. B. 1985. Iron Ore Mining: Assessing Morbidity. *Medical Journal of Australia* 143(2): 68–69.

160. Fishback, P.V. 1986. Workplace Safety during the Progressive Era: Fatal Accidents in Bituminous Coal Mining, 1912–1923. *Explorations in Economic History* 23(3): 269–298.

161. Fleischer, R. L. 1981. A Possible Association between Lung Cancer and Phosphate Mining and Processing. *Health Physics* 41(1): 171–175.

162. Ghose, M. K. and Majee, S. R. 2002. Assessment of the Status of Work Zone Air Environment Due to Opencast Coal Mining. *Environmental Monitoring & Assessment* 77(1): 51–60.

163. Graham, B. L., et al. 1984. Pulmonary function and respiratory symptoms in Potash Workers. *Journal of Occupational Medicine* 26(3): 209–214.

164. Graham, W. G., et al. 2004. Vermont Granite Mortality Study. *Journal of Occupational Medicine and Environmental Health* 46(5): 459–466.

165. Grieve, A. 1988. Oil: Toxicity of Drilling Mud. *Occupational Health* 40(12): 736–739.

166. IARC. 1987. *IARC Monographs on the Evaluation of Carcinogenic Risks to Humans: Silica and Some Silicates* (Monograph No. 11). Lyon, France: IARC.

167. Knox, J. M., et al. 1986. Acquired Perforating Disease in Oil Field Workers. *Journal of the American Academy of Dermatology* 14(4): 605–611.

168. Kucha, H., et al. 1993. Health Risks and Natural Gas. *Nature* 363(6431): 680.

169. Kullman, G. J., et al. 1995. Occupational Exposures to Fibers and Quartz at 19 Crushed Stone Mining and Milling Operations. *American Journal of Industrial Medicine* 27(5): 641–660.

170. Kurilo, S. M. 1998. The Characteristics of the Course of Chronic Dust-induced Bronchitis in Workers in Granite Quarries. *Lik Sprava* 8: 50–53.

171. Landen, D., et al. 2004. Noise Exposure and Hearing Loss among Sand and Gravel Miners. *Journal of Occupational & Environmental Hygiene* 1(8): 532–541.

172. Loomis, D., et al. 1999. Fatal Occupational Injuries among Electric Power Company Workers. *American Journal of Industrial Medicine* 35(3): 302–309.

173. Love, R. G., et al. 1998. Respiratory Health Effects of Opencast Coal Mining. *Occupational and Environmental Medicine* 54: 416–423.

174. Maher, N. 2003. The Application of a Job Exposure Matrix in the Natural Gas Industry. *AIHA J* 64(6): 806–814.

175. Matolo, N. M., et al. 1972. Coal Mining and Cancer of the Stomach. *Rocky Mountain Medical Journal* 69(12): 44–49.

176. McNabb, S. J., et al. 1994. Injuries to International Petroleum Drilling Workers, 1988–1990. *Journal of Occupational Medicine* 36(6): 627–630.

177. Meszaros, G., et al. 2004. Long-Term Persistence of Chromosome Aberrations in Uranium Miners. *Journal of Occupational Health* 46(4): 310–315.

178. Methner, M. M. 2000. Identification of Potential Hazards Associated with New Residential Construction. *Applied Occupational Environmental Hygiene* 15(2): 189–192.

179. Methner, M. M., et al. 2000. Task-Based Exposure Assessment of Hazards Associated with New Residential Construction. *Applied Occupational Environmental Hygiene* 15(11): 811–819.

180. Miller, B. G. and Jacobsen, M. 1985. Dust Exposure, Pneumoconiosis, and Mortality of Coal Miners. *British Journal of Industrial Medicine* 42(11): 723–733.

181. Morgan, W. K. 1999. Coal Mining and COPD. *Thorax* 54(2): 187.

182. Moure-Eraso, R. 1999. Observational Studies as Human Experimentation: The Uranium Mining Experience in the Navajo Nation (1947–1966). *New Solutions* 9(2): 163–178.

183. Ng, T. P. and Chan, S. L. 1992. Lung Function in Relation to Silicosis and Silica Exposure in Granite Workers. *European Respiratory Journal* 5(8): 986–991.

184. NIOSH. 1992. *Preventing Silicosis and Death in Rock Drillers* (Pub. 92–107). Washington, DC: NIOSH.

185. Ogola, J. S. 2002. Impact of Gold Mining on the Environment and Human Health. *Environmental Geochemistry & Health* 24(2): 141–157.

186. Ormerod, A. D., et al. 1998. Novel Causes of Contact Dermatitis from Offshore Oil-based Drilling Muds. *Contact Dermatitis* 39(5): 262–263.

187. Reardon, J. 1993. Injuries and Illness among Bituminous and Lignite Coal Miners. *Monthly Labor Review* 116(10): 49–55.

188. Ruckley, V. A., et al. 1984. Emphysema and Dust Exposure in a Group of Coal Workers. *American Review of Respiratory Disease* 129(4): 528–532.

189. Saverin, R., et al. 1999. Diesel Exhaust and Lung Cancer Mortality in Potash Mining. *American Journal of Industrial Medicine* 36(4): 415–422.

190. Seager, D. R. 2001. Barre, Vermont Granite Workers and the Struggle against Silicosis, 1890–1960. *Labor History* 42(1): 61–79.

191. Sharpe, W. D. 1978. The New Jersey Radium Dial Painters: A Classic in Occupational Carcinogenesis. *Bulletin of the History of Medicine* 52(4): 560–570.

192. Taylor, D. M. and Taylor, S. K. 1997. Environmental Uranium and Human Health. *Review of Environmental Health* 12(3): 147–157.

Sector 22: Utilities

2211: Electric Power Generation, Transmission, and Distribution
2212: Natural Gas Distribution
2213: Water, Sewage, and Other Systems

READINGS IN HISTORY

1. Abell, J. D. and Glumac, P. D. 1997. Beneath the MCI Center: Insights into Washington's Historic Water Supply. *Washington History* 9(1): 24–41.
2. Alexander, T. G. 1968. The Powell Irrigation Survey and the People of the Mountain West. *Journal of the West* 7(1): 48–54.
3. ——. 2003. Interdependence in the Mormon Heartland: Mutual Irrigation Companies and Modernization in Utah's Wasatch Oasis, 1870–1930. *Mining History Journal* 10: 87–102.
4. Arrington, L. J. 1971. The First Irrigation Reservoir in the United States: The Newton, Utah Project. *Utah Historical Quarterly* 39(3): 207–223.
5. ——. 1986. Irrigation in the Snake River Valley: An Historical Overview. *Idaho Yesterdays* 30(1–2): 3–11.
6. Banks, F. E. 2003. An Introduction to the Economics of Natural Gas. *OPEC Review* 27(1): 25.
7. Bhansali, A., et al. 1992. Sewage Treatment as an Industrial Subsidy. *Economic Geography* 68(2): 174–187.

8. Bogener, S. D. 1998. Ditches across the Desert: A Story of Irrigation along New Mexico's Pecos River. Dissertation, Texas Tech University.

9. Bott, D. M. 1997. Mains, Drains and Strains: The Emergence of Urban Water and Waste Systems in the United States. Dissertation, University of Illinois, Urbana-Champaign.

10. Britton, C. C. 1972. An Early Electric Power Facility in Colorado. *Colorado Magazine* 49(3): 185–195.

11. Brown, C. D. 1980. *Electricity for Rural America: The Fight for the REA.* Westport, CT: Greenwood Press.

12. Cannon, B. Q. 1992. Quite a Wrestling Match: The Adaptation of Dry Land Farmers to Irrigation. *Agricultural History* 66(2): 120–136.

13. Clarfield, G. H. and Wiecek, W. M. 1984. *Nuclear America: Military and Civilian Nuclear Power in the United States, 1940–1980.* New York: Harper & Row.

14. Cotroneo, R. R. 1992. Schemes and Dreams: Irrigation Projects on the Northern Pacific Land Grant. *Pacific Northwest Forum* 5(2): 37–50.

15. Cowan, R. 1990. Nuclear Power Reactors: A Study in Technological Lock-in. *Journal of Economic History* 50(3): 541–567.

16. Creutz, E. 1970. Nuclear Power: Rise of an Industry. *Bulletin of the Atomic Scientists* 26(6): 75–82.

17. Cruickshank, A. B. 1986. Commercial Nuclear Power Plants in the USA. *Geography* 71(1): 73–76.

18. Dahl, C. A. and Matson, T. K. 1998. Evolution of the U.S. Natural Gas Industry in Response to Changes in Transaction Costs. *Land Economics* 74(3): 390–408.

19. Davidson, J. E. 2000. Power Switches on the Middle Snake River: The Divergent Histories of Two Hydroelectric Projects. *Idaho Yesterdays* 44(2): 22–32.

20. Dawson, F. G. 1976. *Nuclear Power: Development and Management of a Technology.* Seattle: University of Washington Press.

21. Doran, W. A. 1968. Early Hydroelectric Power in Tennessee. *Tennessee Historical Quarterly* 27(1): 72–82.

22. DuBoff, R. B. 1979. *Electric Power in American Manufacturing, 1889–1958.* New York: Arno.

23. Dunbar, W. P. 1908. *Principles of Sewage Treatment.* Philadelphia: J.B. Lippincott Company.

24. Ellis, J. and Galishoff, S. 1977. Atlanta's Water Supply 1865–1918. *Maryland Historian* 8(1): 5–22.

25. Emmons, W. M. 1993. Franklin D. Roosevelt, Electric Utilities and the Power of Competition. *Journal of Economic History* 53(4): 880–907.

26. Ford, J. 1982. Electricity for a Region: The Southwestern Power Administration. *Chronicles of Oklahoma* 60(4): 448–459.

27. Foss-Mollan, K. 2001. *Hard Water: Politics and Water Supply in Milwaukee, 1870–1995*. West Lafayette, IN: Purdue University Press.

28. Gandy, M. 1997. The Making of a Regulatory Crisis: Restructuring New York City's Water Supply. *Transactions of the Institute of British Geographers* 22(3): 338.

29. Goldman, J. A. 1997. Building New York's Sewers: Developing Mechanism of Urban Management. Dissertation, Purdue University.

30. Gracy, D. B. 1984. Irrigation on the High Plains—A Perspective. *Red River Valley Historical Review* 1(2): 127–131.

31. Green, B. D. 1986. Irrigation Expansion in Arkansas: A Preliminary Investigation. *Arkansas Historical Quarterly* 45(3): 261–268.

32. Green, D. E. 1973. *Land of the Underground Rain: Irrigation on the Texas High Plains, 1910–1970*. Austin: University of Texas Press.

33. Gregory, G. P. 1974. A Study in Local Decision Making: Pittsburgh and Sewage Treatment. *Western Pennsylvania Historical Magazine* 57(1): 25–42.

34. Greenberg, M. R., et al. 1971. A Geographical Systems Analysis of the Water Supply Networks of the New York Metropolitan Region. *Geographical Review* 61(3): 339–354.

35. Heppenheimer, T. A. 2002. Nuclear Power: What Went Wrong? *American Heritage of Invention & Technology* 18(2): 46–50, 52–56.

36. Hirsh, R. F. 1999. *Power Loss: The Origins of Deregulation and Restructuring in the American Electric Utility System*. Cambridge, MA: MIT Press.

37. ———. 2003. *Technology and Transformation in the American Electric Utility Industry*. Cambridge: Cambridge University Press.

38. Hope, B. and Sheehan, M. 1983. The Political Economy of Centralized Water Supply in California. *Social Science Journal* 20(2): 29–39.

39. Howard, S. W. 1992. *Green Fields of Montana: A Brief History of Irrigation*. Manhattan, KS: Sunflower University Press.

40. Hurt, R. D. 1979. Irrigation in the Kansas Plains Since 1930. *Red River Valley Historical Review* 4(3): 64–72.

41. ——. 1991. Irrigation in the West. *Journal of the West* 30(2): 63–77.

42. Kepfield, S. S. 1994. El Dorado on the Platte: The Development of Agricultural Irrigation and Water Law in Nebraska, 1860–1895. *Nebraska History* 75(3): 232–243.

43. ——. 1995. Great Plains Legal Culture and Irrigation Development. *Environmental History Review* 19(4): 49–66.

44. Lee, K. N. 1981. Nuclear Power and Electrical Energy. *Policy Studies Journal* 9(7): 1087–1092.

45. Lilley, W. and Lewis, L. 1996. The Western Irrigation Movement, 1878–1902: A Reappraisal. *University of Wyoming Publications* 32: 57–74, 152–157.

46. Lovin, H. T. 2002. Dreamers, Schemers, and Doers of Idaho Irrigation. *Agricultural History* 76(2): 232–243.

47. Luxton, P. 1998. And Then There Was Light: The History of Rural Electricity in West Texas. *Permian Historical Annual* 38: 73–88.

48. Marvin, A. M. 1996. A Grave-yard of Hopes: Irrigation and Boosterism in Southwest Kansas, 1880–1890. *Kansas History* 19(1): 36–51.

49. McCormick, J. S. 1988. The Beginning of Modern Electric Power Service in Utah, 1912–1922. *Utah Historical Quarterly* 56(1): 4–22.

50. McDevitt, E. P. 1995. The Evolution of Irrigation Institutions in California: The Rise of the Irrigation District, 1910–1930. Dissertation, University of California, Los Angeles.

51. McVicker, C. H. 1985. The Changing Spatial Distribution of Electric Generation in New Jersey. Dissertation, Rutgers University, New Brunswick.

52. Murray, K. A. 1996. The Highline Canal: Irrigation Comes to Wenatchee. *Columbian* 9(4): 17–23.

53. Nesson, F. L. 1983. *Great Waters: A History of Boston's Water Supply*. Hanover, NH: University Press of New England.

54. Norby, O. and Higham, R. 1983. Irrigation in Western Kansas—The 1950s. *Journal of the West* 22(2): 26–29.

55. Nye, E. E. 1990. *Electrifying America: Social Meanings of a New Technology, 1880–1940*. Cambridge, MA: MIT Press.

56. O'Dell, G. A. 1993. Water Supply and the Early Development of Lexington, Kentucky. *Filson Club History Quarterly* 67(4): 431–461.

57. Ogle, M. 1999. Water Supply, Waste Disposal, and the Culture of Privatism in the Mid 19th Century American City. *Journal of Urban History* 25(3): 321–347.

58. Pierce, M. A. 1993. The Introduction of Direct Pressure Water Supply, Cogeneration, and District Heating in Urban and Institutional Communities, 1863–1882. Dissertation, University of Rochester.

59. Pisani, D. J. 1984. *From the Family to Agribusiness: The Irrigation Crusade in California and the West, 1850–1931*. Berkeley: University of California Press.

60. Relander, C. 1961. The Battleground of National Irrigation. *Pacific Northwest Quarterly* 52(4): 144–150.

61. Rose, N. L. and Joskow, R. L. 1990. The Diffusion of New Technologies: Evidence from the Electric Utility Industry. *The RAND Journal of Economics* 21(3): 354–373.

62. Rossi, M. C. 1995. The History of Sewage Treatment in the City of Buffalo, NY. *Middle States Geographer* 28: 9–19.

63. Rowland, M. A. 2001. The Evolution of Water Resource Management Systems. Dissertation, University of New Orleans.

64. Schap, D. J. 1983. A History of Municipal Ownership in the Electric Power Industry in the United States: A Centennial View. Dissertation, Washington University.

65. Schurr, S. H., et al. 1990. *Electricity in the American Economy: Agent of Technological Progress*. Westport, CT: Greenwood Press.

66. Sholders, M. 2002. Water Supply Development in San Diego and a Review of Related Outstanding Projects. *Journal of San Diego History* 48(1): 60–71.

67. Sidwick, J. M. 1977. *A Brief History of Sewage Treatment*. Middlesex, UK: Thunderbird Enterprises.

68. Smith, B. A. 1974. Technological Innovation in Electric Power Generation, 1950–1970. Dissertation, Indiana University.

69. Sommers, P. E. 1979. The Diffusion of Nuclear Power Generation in the United States. Dissertation, Yale University.

70. Tomayko, J. E. 1981. Irrigation Technology and the Development of Southwest Kansas. Dissertation, Carnegie-Mellon University.

71. Troesken, W. 1996. *Why Regulate Utilities: The New Institutional Economics and the Chicago Gas Industry, 1849–1924*. Ann Arbor: University of Michigan Press.

72. Zeisler-Vralsted, D. 1988. A History of the Kennewick Irrigation District, State of Washington, 1880–1987. Dissertation, Washington State University.

READINGS IN HAZARDOUS WASTE MANAGEMENT AND POLLUTION PREVENTION

73. EPRI. 1987. *Inorganic and Organic Constituents in Fossil Fuel Combustion Residue, Volume 1: A Critical Review* (EP-5176).

74. Geldreich, E. E. 1996. *Microbial Quality of Water Supply in Distribution Systems.* Boca Raton, FL: CRC Press.

75. Henry, C., et al. 1999. Managing Nitrogen from Biosolids (99–508). A Report Prepared for the Washington Department of Ecology.

76. Pacific Northwest Pollution Prevention Research Center. 1993. *Pollution Prevention Opportunities in Oil and Gas Production, Drilling and Exploration*. Seattle, PPRC.

77. University of Louisville. 2002. *Energy Efficiency for Wastewater Operations*. Louisville, Kentucky Pollution Prevention Center.

78. U.S. Department of the Interior. *Pollution Prevention Handbook: Sewage & Wastewater Treatment Plants* (No. 13). Washington, DC: Office of Environmental Affairs.

79. U.S. Environmental Protection Agency. 1985. *Guidance Manual for Implementing Total Toxic Organics Pretreatment Standards* (EPA/440/1–85/009T). Darby, PA: Diane Publishing.

80. ——. 1987. *Guidance Manual for Preventing Interference at POTWs*. Washington, DC: Office of Water Enforcement & Permits.

81. ——. 1988. *Wastes from the Combustion of Coal by Electric Utility Power Plants* (EPA/530/SW-88/002). Office of Solid Waste & Emergency Response.

82. ——. 1993. *Guides to Pollution Prevention: Municipal Pretreatment Programs* (EPA/625/R-93/006). Washington, DC: EPA.

83. ——. 1997. *Profile of the Fossil Fuel Electric Power Generation Industry* (EPA/310/R-97/007). Washington, DC: EPA.

84. ——. 1999a. *Wastes from the Combustion of Fossil Fuels, Volume 1—Executive Summary* (EPA/530/S-99/010). Washington, DC: Office of Solid Waste & Emergency Response.

85. ——. 1999b. *Wastes from the Combustion of Fossil Fuels, Volume 2—Methods, Findings & Recommendations* (EPA/530/R-99/010). Washington, DC: Office of Solid Waste & Emergency Response.

86. Young, L. B. and Young, H. P. 1984. Pollution by Electrical Transmission: The Environmental Impact of High Voltage Lines. *Bulletin of Atomic Scientists* 30(10): 34–38.

READINGS IN HEALTH AND SAFETY

87. Bonnell, J. A. 1982. Effects of Electric Fields Near Power Transmission Plants. *Journal of the Royal Society of Medicine* 75: 933.

88. Cohen, A. A., et al. 1972. Asthma and Air Pollution from a Coal-fueled Power Plant. *American Journal of Public Health* 62: 1181–1188.

89. Douwes, J., et al. 2001. Work-Related Symptoms in Sewage Treatment Workers. *Annals of Agricultural & Environmental Medicine* 8: 39–45.

90. Driscoll, T., et al. 1995. Health Status and Exposure of Workers at a Pilot Brown Coal Liquefaction Plant in Australia, 1985–1991. *Occupational Medicine* 45(5): 239–246.

91. Hampt, R. C. and Nolfi, V. R. 1984. Health Effect of High Voltage Transmission Lines on the Health of Adjacent Resident Population. *American Journal of Public Health* 74: 76.

92. Kindzierski, W. B. 1997. Health Effects Associated with Wastewater Treatment, Disposal and Reuse. *Water Environment Research* 69(4): 915–925.

93. Lundholm, M. and Rylander, R. 1983. Work-Related Symptoms among Sewage Workers. *British Journal of Industrial Medicine* 40: 325–329.

94. Mattsby, I. and Rylander, R. 1978. Clinical and Immunological Findings in Workers Exposed to Sewage Dust. *Journal of Occupational Medicine* 20(10): 690–692.

95. McCunney, R. J. 1986. Health Effects of Work at Wastewater Treatment Plants. *American Journal of Industrial Medicine* 9: 271–279.

96. McDowall, M. E. 1986. Mortality of Persons Resident in the Vicinity of Electricity Transmission Facilities. *British Journal of Cancer* 53: 271–279.

97. Neu, B. A. 1979. Avoiding Confined Space Hazards in Wastewater Treatment Plants. *Pollution Engineering* (August).

98. Petrelli, G., et al. 1989. A Retrospective Cohort Mortality Study on Workers of Two Thermoelectric Power Plants: Fourteen-year Follow-up Results. *European Journal of Epidemiology* 5(1): 87–89.

99. Prazmo, Z., et al. 2003. Exposure to Bioaerosols in a Municipal Sewage Treatment Plant. *Annals of Agricultural & Environmental Medicine* 10: 241–248.

100. Zuskin E., et al. 1993. Respiratory Function in Sewage Workers. *American Journal of Industrial Medicine* 23: 751–761.

Sector 23: Construction

233: Building, Developing, and General Contracting
234: Heavy Construction
235: Special Trade Contractors

READINGS IN HISTORY

1. Adams, D. R. 1975. Residential Construction Industry in the Early 19th Century. *Journal of Economic History* 35(4): 794–816.
2. Alberts, W. W. 1962. Business Cycles, Residential Construction, and the Mortgage Market. *Journal of Political Economy* 70(3): 263–281.
3. Aldrich, M. 1999. Engineering Success and Disaster: American Railroad Bridges, 1840–1900. *Railroad History* 180: 31–72.
4. Allen, W. J. 1971. The Multifamily Housing Solution and Housing Type Preferences. *Social Science Quarterly* 52(3): 543–559.
5. Axelrod, A. 1979. A Century of Light: The Development of Iowa's Electric Utilities. *Palimpsest* 60(5): 130–131.
6. Ball, R. 1973. Labor and Materials Required for Highway Construction. *Monthly Labor Review* 96(6): 40–45.
7. Barr, K. 1992. The National Movement to Halt the Spread of Multifamily Housing, 1890–1926. *Journal of the American Planning Association* 58(1): 39–48.
8. Barton, E. M. 1962. The Covered Bridges of Columbia County, Pennsylvania. *Columbian* 1(3): 3–32.

9. Beach, U. S. 1969. Tennessee's Covered Bridges. *Tennessee Historical Quarterly* 28 (1): 3–23.

10. Bender, B. 1979. The Determinants of Housing Demolition and Abandonment. *Southern Economic Journal* 46(1): 131–144.

11. Bromhead, C. E. N. 1942. The Early History of Water Supply. *The Geographical Journal* 99(3): 142–151.

12. Brucemann, R. and Prowler, D. 1977. 19th Century Mechanical System Designs. *JAE* 30(3): 11–15.

13. Buckley, T. C. 2001. Trucks along the Southwest Border: Army Motorization and Highway Construction in the U.S.–Mexico Border Country, 1916–1926. *Journal of Big Bend Studies* 13: 215–234.

14. Caldwell, D. J. 1967. Missouri's Covered Bridges. *Missouri Historical Review* 61(2): 229–236.

15. Castaneda, C. J. 1990. The Texas–Northeast Connection: The Rise of the Post–World War II Gas Pipeline Industry. *Houston Review* 12(2): 79–95.

16. Cavanaugh, E. H. 2003. Who Designed Your House? A Technological and Cultural History of Conventional Wood Construction, 1790–1880. Dissertation, Lehigh University.

17. Clemhout, S. 1981. The Impact of Housing Cyclicality on the Construction of Residential Units and Housing Costs. *Land Economics* 57(4): 609–623.

18. Coates, P. A. 1992. *The Trans-Alaska Pipeline Controversy: Technology, Conservation, and the Frontier*. Bethlehem, PA: Lehigh University Press.

19. Coffey, B. and Noble, A. G. 1989. Residential Building Materials in New York State, 1855–1875. *Material Culture* 21(1): 3–21.

20. Cheever, L. O. 1970. Census of Covered Bridges. *Palimpsest* 51(11): 466–496.

21. Cooper, G. 2002. *Air Conditioning America: Engineers and the Controlled Environment, 1900–1960*. Baltimore, MD: Johns Hopkins University Press.

22. Cooper, J. L. 1987. *Iron Monuments to Distant Prosperity: Indiana's Metal Bridges 1870–1930.* Greencastle, IN: DePauw University.

23. Danko, G. M. 1979. The Evolution of the Simple Truss Bridge 1790 to 1850. Dissertation, University of Pennsylvania.

24. DeLony, E. 1993. Surviving Cast and Wrought Iron Bridges in America. *Journal of the Society for Industrial Archaeology* 19(2): 17–47.

25. Dey, P. K. 2001. A Risk-based Model for Inspection and Maintenance of Cross-country Petroleum Pipelines. *Journal of Quality in Maintenance Engineering* 7(1): 25–43.

26. Draffin, J. O. 1943. *A Brief History of Lime, Cement, Concrete and Reinforced Concrete*. Urbana: University of Illinois Press.

27. Evans, R. E. 1977. Highways to Progress: 19th Century Roads in the Cherokee Nation. *Journal of Cherokee Studies* 2(4): 394–400.

28. Fitzsimons, G. 1996. Uncivil Engineers: The Struggle for Control of Seattle's Early Water and Electric Utilities, 1890–1910. *Journal of the Society for Industrial Archaeology* 22(1): 10–34.

29. Foster, H. G. 1972. Unions, Residential Construction, and Public Policy. *Quarterly Review of Economics & Business* 12(4): 45–56.

30. Fraser, C. B. 1985. Bridging History: Historic Bridges of Colorado. *Colorado Heritage* 3: 16–32.

31. Goldman, J. A. 1997. Building New York's Sewers: Developing Mechanisms of Urban Management. Dissertation, Purdue University.

32. Gottfried, H. 1995. The Machine and the Cottage: Building, Technology, and the Single-Family House, 1870–1910. *The Journal of the Society for Industrial Archeology* 21(2): 47–68.

33. Greene, J. W. 2005. Sources for the History of the Building and Construction Industry. *Labor History* 46(4): 495–511.

34. Gruber, J. 1981. Wisconsin River Railroad Bridges: A Historic Resource. *Pioneer America Society Transactions* 4: 55–65.

35. Hadlow, R. W. 1994. Bridges over Washington: A Historic Engineering Record. *Columbia* 8(1): 6–11.

36. Harris, L. J. 1999. Wrecking to Save: The Chicago House Wrecking Company. *Journal of the West* 38(4): 65–74.

37. Holley, I. B. 2003. Blacktop: How Asphalt Paving Came to the Urban United States. *Technology and Culture* 44(4): 703–733.

38. Howard, G. W. 1976. Bridges in the Desert: Early Days of the Yuma Proving Ground. *Journal of Arizona History* 17(4): 431–450.

39. Hyde, C. K. 1993. *Historic Highway Bridges of Michigan*. Detroit: Wayne State University Press.

40. Jakle, J. A. 2000. Pioneer Roads: America's Early 20th Century Named Highways. *Material Culture* 32(2): 1–22.

41. Johnson, L. R. 1978. Memphis to Bristol: A Half Century of Highway Construction: A History of The Tennessee Road Builders Association, 1928–1978. Privately published.

42. Kaszynski, W. 2000. *The American Highway: The History and Culture of Roads in the United States*. Jefferson, NC: McFarland & Company.

43. Keating, A. D. 1994. *Invisible Networks: Exploring the History of Local Utilities and Public Works*. Malabar, FL: Krieger.

44. Knerr, D. 2004. *Suburban Steel: The Magnificent Failure of the Lustron Corporation, 1945–1951*. Columbus: Ohio State University Press.

45. Kniffen, F. 1951. The American Covered Bridge. *Geographical Review* 41(1): 114–123.

46. Kranakis, E. 1997. *Constructing a Bridge: An Exploration of Engineering Culture, Design & Research in 19th Century France and America*. Cambridge, MA: MIT Press.

47. Kroessler, J. A. 1989. Bridges and the Urban Landscape. *Long Island Historical Journal* 2(1): 104–117.

48. Lamb, B. P. 1996. Highways and Waterways: Two Episodes in California's Transportation History. *California History* 75(1): 12–20, 105.

49. Lewis, T. 1997. *Divided Highways: Building the Interstate Highways, Transforming American Life*. New York: Viking Penguin.

50. Longstreth, R. 1987. *The Buildings of Main Street: A Guide to American Commercial Architecture*. Walnut Creek, CA: AltaMira.

51. MacKenzie, C. L. 2003. House Building on the Vineyard between the Two World Wars. *Dukes County Intelligencer* 45(1): 3–12.

52. Marzio, P. C. 1972. Carpentry in the Southern Colonies during the 18th Century. *Winterthur Portfolio* 7: 229–250.

53. Massey, R. C. and Ewig, R. 1986. Wyoming's Truss Bridges. *Annals of Wyoming* 58(2): 46–53.

54. McCroskey, L. L. 1992. Historical Architecture. Artful Passages: The Engineering Legacy of North Dakota Bridges. *North Dakota History* 59(2): 28–32.

55. Mikesell, S. D. 1986. The Los Angeles River Bridges. A Study in the Bridge as a Civic Monument. *Southern California Quarterly* 68(4): 365–386.

56. Miller, M. R. and Miller, R. 2005. *Miller's Guide to Foundations and Sitework*. New York: McGraw-Hill Professional.

57. Nelson, J. C. 1997. *Spanning Time: Vermont's Covered Bridges*. Shelburne, VT: New England Press.

58. Nelson, L. H. 1960. A Century of Oregon Covered Bridges. *Oregon Historical Quarterly* 61(2): 101–209.

59. O'Brien, J. J., et al. 1996. *Standard Handbook of Heavy Construction*. New York: McGraw Hill.

60. O'Flaherty, C. A. 2002. *Highways: The Location, Design, Construction & Maintenance of Road Pavements*. New York: Elsevier.

61. Ogle, M. 1996. *All the Modern Conveniences: American Household Plumbing, 1840–1890.* Baltimore, MD: Johns Hopkins University Press.

62. Ott, K. 1999. All the Modern Conveniences: American Household Plumbing, 1840–1890. *Bulletin of the History of Medicine* 73(2): 336–338.

63. Palladino, G. 2005. *Skilled Hands, Strong Spirits: A Century of Building Trades History*. Ithaca, NY: Cornell University Press.

64. Petersen, J. G. 2001. The Lincoln Highway and Its Changing Routes in Utah. *Utah Historical Quarterly* 69(3): 192–214.

65. Peterson, C. E. 1969. The Technology of Early American Building. *Newsletter of the Association for Preservation Technology* 1(1): 3–17.

66. Pierce, M. A. 1993. The Introduction of Direct Pressure Water Supply, Cogeneration, and District Heating in Urban and Institutional Communities, 1863–1882. Dissertation, University of Rochester.

67. Pillsbury, R. 1976. The Construction Materials of the Rural Folk Housing of the Pennsylvania Culture Region. *Pioneer America* 8(2): 98–106.

68. Plowden, D. 1974. *Bridges: The Spans of North America*. New York: Viking.

69. Potter, J. E. and Puschendorf, R.L., eds. 1999. *Spans in Time: A History of Nebraska Bridges.* Lincoln: Nebraska State Historical Society.

70. Prier, R. J. 1979. Labor and Material Requirements for Federally Aided Highways. *Monthly Labor Review* 102(12): 29–34.

71. Randall, F. A. 1999. *History of the Development of Building Construction in Chicago*. Chicago: University of Illinois Press.

72. Raub, P. 1990. Another Pattern of Urban Living: Multifamily Housing in Providence, 1890–1930. *Rhode Island History* 48(1): 3–19.

73. Reyes, G. and Rich, P. 2003. Housing Students: Fraternities and Residential Colleges. *Annals of the American Academy of Political & Society Science* 585: 118–123.

74. Ritchie, T. 1973. Notes on the History of Hollow Masonry Walls. *Bulletin of the Association for Preservation Technology* 5(4): 40–49.

75. Rome, A. W. 1994. Building on the Land: Toward an Environmental History of Residential Development in American Cities and Suburbs, 1870–1990. *Journal of Urban History* 20(3): 407–434.

76. Rose, W. B. 1997. Moisture Control in the Modern Building Envelope: History of the Vapor Barrier in the U.S., 1923–1952. *APT Bulletin* 28(4): 13–19.

77. Rowland, M. 1982. Kansas and the Highways, 1917–1930. *Kansas History* 5(1): 33–51.

78. Schmitt, R. C. 1981. Some Construction and Housing Firsts in Hawaii. *Hawaiian Journal of History* 15: 100–112.

79. ———. 1982. Pipes, Pools, and Privies: Some Notes on Early Island Plumbing. *Hawaiian Journal of History* 16: 149–170.

80. ———. 1986. Early Hawaiian Bridges. *Hawaiian Journal of History* 20: 151–157.

81. Schultz, S. K. and McShane, C. 1978. To Engineer the Metropolis: Sewers, Sanitation, and City Planning in Late 19th Century America. *Journal of American History* 65(2): 398–411.

82. Seeley, B. E. 1987. *Building the American Highway System: Engineers and Policy Makers*. Philadelphia: Temple University Press.

83. Simmons, D. A. 1985. Engineering and Enterprise: Early Metal-truss Bridges in Ohio. *Timeline* 2(1): 14–29.

84. ———. 1993. Bridges and Boilers: Americans Discover the Wrought-iron Tubular Bowstring Bridge. *Journal of the Society for Industrial Archeology* 19(2): 63–76.

85. Skidmore, M. J. 1997. The Theodore Roosevelt International Highway. *Theodore Roosevelt Association Journal* 22(1–2): 15–19.

86. Smith, H. 1980. Covered Bridges: Sturdy Relics of Missouri. *Missouri Historical Review* 74(2): 123–136.

87. Slaton, A. E. 2001. *Reinforced Concrete and the Modernization of American Building, 1900–1930.* Baltimore, MD: Johns Hopkins University Press.

88. Speweik, J. P. 1995. *History of Masonry Mortar in America, 1720–1995.* Arlington, VA: National Lime Association.

89. Stanley, C. C. 1979. *Highlights in the History of Concrete*. Crowthorne, UK: British Cement Association.

90. Stone, M. N. 1979. The Plumbing Paradox: American Attitudes toward Late 19th Century Domestic Sanitary Arrangements. *Winterthur Portfolio* 14(3): 283–309.

91. Strawbridge, L. L. 1989. The Walnut Lane Bridge: First Prestressed Concrete Bridge in the U.S. *Germantown Crier* 41(2): 28–33.

92. Swiatosz, S. 1985. A Technical History of Late 19th Century Windows in the United States. *APT Bulletin* 17(1): 31–37.

93. Troyano, L. F. 2004. *Bridge Engineering: A Global Perspective*. Reston, VA: ASCE.

94. Twitchell H. 1992. *Northwest Epic: The Building of the Alaska Highway*. New York: St. Martin's Press.

95. Walter, J. E. 2000. Highway Construction over the Summit: The Cheyenne Pass Telephone Canyon Road. *Annals of Wyoming* 72(2): 30–34.

96. Watson, S. R. and Wolfs, J. R. 1981. Bridges of Metropolitan Cleveland: Past and Present. Cleveland: Privately published.

97. Weiss, N. R. 1995. Chemical Treatments for Masonry: An American History. *APT Bulletin* 26(4): 9–16.

98. Woodward, C. M. 1881. *A History of the St. Louis Bridge: Containing a Full Account of Every Step in Its Construction and Erection*. St. Louis: G.I. Jones & Co.

99. Wyckoff, W. 1986. Land Subdivision on the Holland Purchase in Western New York State, 1797–1820. *Journal of Historical Geography* 12(2): 142–161.

100. Yearwood, R. M. 1970. Land Subdivisions and Development: American Attitudes on Land Subdivision and Its Controls. *American Journal of Economics and Sociology* 29(2): 113–126.

101. Yeomans, D. T. 1991. British and American Solutions to a Roofing Problem. *Journal of the Society of Architectural Historians* 50(3): 266–272.

READINGS IN HAZARDOUS WASTE MANAGEMENT & POLLUTION PREVENTION

102. Ball, R. 1973. Labor and Materials Required for Highway Construction. *Monthly Labor Review* 96(6): 40–45.

103. Klang, A., et al. 2003. Sustainable Management of Demolition Waste. *Resources, Conservation and Recycling* 38(4): 317–334

104. Mroueh, U-M., et al. 2000. *Life Cycle Assessment of Road Construction.* Helsinki: Finnish National Road Administration.

105. U.S. Environmental Protection Agency. 1998. *Characterization of Building-Related Construction and Demolition Debris in the United States* (EPA/530/R-98/010). Washington, DC: Office of Solid Waste & Emergency Response.

106. ——. 2004. *RCRA in Focus: Construction, Demolition and Renovation* (EPA/530/K-04/005). Washington, DC: Office of Solid Waste & Emergency Response.

READINGS IN HEALTH AND SAFETY

107. Baker, E. L., et al. 1988. Neurobehavioral Effects of Solvents in Construction Painters. *Journal of Occupational Medicine* 30: 116–123.

108. Bakke, B., et al. 2004. Cumulative Exposure to Dust and Gases as Determinants of Lung Function Decline in Tunnel Construction Workers. *Occupational and Environmental Medicine* 61: 262–269.

109. Burns, C., et al. 1962. Health Hazards and Heavy Construction. *AIHA Journal* 23: 273–281.

110. Burstyn, I., et al. 2000. Literature Review of Levels and Determinants of Exposure to Potential Carcinogens and Other Agents in Road Construction Industry. *AIHA Journal* 61: 715–726.

111. Cantor, K. P., et al. 1986. Patterns of Mortality among Plumbers and Pipe Fitters. *American Journal of Industrial Medicine* 10(1): 73–89.

112. Fidler, A. T., et al. 1987. Neurobehavioral Effects of Occupational Exposure to Organic Solvents among Construction Painters. *British Journal of Industrial Medicine* 44: 292–304.

113. Flynn, M. R. and Susi, P. 2003. Engineering Controls for Selected Silica and Dust Exposures in the Construction Industry—A Review. *Applied Occupational Environmental Hygiene* 18(4): 268–277.

114. Gerwel, B., et al. 1993. Lead Exposure in Bridge Construction Workers. *American Journal of Public Health* 83(7): 1054–1055.

115. Kaminski, R., et al. 1980. Mortality Analysis of Plumbers and Pipe Fitters. *Journal of Occupational Medicine* 22(3): 183–189.

116. Lipscomb, H. J., et al. 2000. Work-Related Injuries in Drywall Installation. *Applied Occupational Environmental Hygiene* 15(10): 794–802.

117. ———. 2003. Work-Related Injuries in Residential and Drywall Carpentry. *Applied Occupational Environmental Hygiene* 18(6): 479–488.

118. Lowry, S. 1992. Injuries from Domestic Glazing. *British Medical Journal* 304(6823): 332.

119. McDowall, M. 1983. Leukemia Mortality in Electrical Workers in England and Wales. *Lancet* 1(8318):246.

120. Moir, S., et al. 2003. Making Sense of Highway Construction: A Taxonomic Framework for Ergonomic Exposure Assessment and Intervention Research. *Applied Occupational Environmental Hygiene* 18(4): 256–267.

121. Neubert, F. R. 1968. Nail Gun and Masonry Nail Accidents. *British Medical Journal* 1(590): 511.

122. NIOSH. 1985. *Preventing Deaths and Injuries form Excavation Cave-ins* (Pub. 85–110). Washington, DC: NIOSH.

123. OSHA. 1998. *Concrete and Masonry Construction* (OSHA 3106). http://www.osha.gov/Publications/3106.html (accessed 15 October 2008).

124. Personick, M. E. 1990. Profiles in Safety and Health Roofing and Sheet Metal Work. *Monthly Labor Review* 113(9): 27–32.

125. Peterson, C. E. 1965. Notes on Copper Roofing in America to 1802. *Journal of the Society of Architectural Historians* 24(4): 313–318.

126. Ringen, K., et al. 1995. Safety and Health in the Construction Industry. *Annual Review of Public Health* 16: 165–188.

127. Roodman, D. M., et al. 1995. *A Building Revolution: How Ecology and Health Concerns Are Transforming Construction*. Washington, DC: Worldwatch Institute.

128. Ruhl, R. and Kluger, N. 1995. Hazardous Substances in Construction Work. *Occupational Medicine* 10(2): 335–351.

129. Shah, S. M., et al. 2003. Injuries and Illness from Wood Framing in Residential Construction, Washington State, 1993–1999. *Journal of Occupational & Environmental Medicine* 45(11): 1171–1182.

130. Stenlund, B., et al. 2002. Significance of House Painter's Work Techniques on Shoulder Muscle Strain during Overhead Work. *Ergonomics* 45(6): 455–468.

131. Van Wijngaarden, E. 2002. Suicide Mortality among Electricians. *Occupational & Environmental Medicine* 59(9): 649.

132. Verma, D. K., et al. 2003. Current Chemical Exposures among Ontario Construction Workers. *Applied Occupational Environmental Hygiene* 18: 1031–1047.

133. Woskie, S. R., et al. 2002. Exposures to Quartz Dust, Diesel and Welding Fumes During Heavy and Highway Construction. *AIHA Journal* 63(4): 447–457.

134. Yereb, D. J. 2003. Evaluation of Dry and Wet Block Cutting and Recommendation for a Masonry Company. *Applied Occupational Environmental Hygiene* 18(3): 145–150.

Sector 31–33: Manufacturing

311: Food Manufacturing
3111: Animal Food Manufacturing
3112: Grain and Oilseed Milling
3113: Sugar and Confectionery Product Manufacturing
3114: Fruit and Vegetable Preserving and Specialty Food Manufacturing
3115: Dairy Product Manufacturing
3116: Animal Slaughtering and Processing
3117: Seafood Product Preparation and Packaging
3118: Bakeries and Tortilla Manufacturing
3119: Other Food Manufacturing

READINGS IN HISTORY

1. Ahmed, Z. Z. and Sieling, M. 1987. Two Decades of Productivity Growth in Poultry Dressing and Processing. *Monthly Labor Review* 110(4): 34–39.

2. Betler, B. 1994. Barg Kass: Cheese Making among the West Virginia Swiss. *Goldenseal* 20(1): 28–35.

3. Blakely, J. R. 2003. The Nestle Condensory in Brandon. *Oregon Historical Quarterly* 104(4): 566–577.

4. Boyd, W. 2001. Making Meat: Science, Technology and American Poultry Production. *Technology and Culture* 42(4): 631–664.

5. Brenner, J. G. 1999. *The Emperors of Chocolate: Inside the Secret World of Hershey and Mars*. New York: Random House.

6. Burroughs, J. M. 1978. New Mexico Peanut Butter: The Real Story. *Palacio* 84(1): 39–41.

7. Clark, V. S. 1916. *History of Manufacturers in the United States*. Washington, DC: Carnegie Institution of Washington.

8. Clifton, J. M. 1978. Charles Manigault's Essay on the Economics of Milling Rice. *Agricultural History* 52(1): 104–110.

9. Colwell, J. L. 1969. From Stone to Steel: American Contributions to the Revolution in Flour Milling. *Rocky Mountain Social Science Journal* 6(2): 20–31.

10. Danbom, D. B. 2003. Flour Power: The Significance of Flour Milling at the Falls. *Minnesota History* 58(5–6): 270–285.

11. Davidson, H. and Snow, S. 1993. The Material Culture of Iowa's Creamery Industry, 1872–1903. *Pioneer America Society Transactions* 16: 25–35.

12. Dupre, R. 1999. If It's Yellow It Must Be Butter: Margarine Regulation in North America since 1886. *Journal of Economic History* 59(2): 353–371.

13. Durrenberger, E. P. 1992. *It's All Politics: South Alabama's Seafood Industry*. Urbana: University of Illinois Press.

14. Ferleger, L. 1984. Cutting the Cane: Harvesting in the Louisiana Sugar Cane Industry. *Southern Studies* 23(1): 42–59.

15. Frazee, P. J. 1994. Alfred Gonzenbach and the Making of the Valley Queen Cheese Factory of Milbank, South Dakota. *Swiss American Historical Society Review* 30(1): 3–25.

16. Frederick, P. B. 2002. *Canning Gold: Northern New England's Sweet Corn Industry: A Historical Geography*. Lanham, MD: University Press of America.

17. Frey, C. N. 1930. History and Development of the Modern Yeast Industry. *Industrial & Engineering Chemistry* 22(11): 1154–1162.

18. Fugate, F. 1994. *Arbuckles: The Coffee That Won the West*. El Paso: Texas Western Press.

19. Graziano, M. M. 1998. Food of the Gods as Mortal's Medicine: The Uses of Chocolate and Cacao Products. *Pharmacy in History* 40(4): 132–146.

20. Gregory, R. W. 1968. Those Good Peas: The Morgan Canning Company in Smithfield, Utah. *Utah Historical Quarterly* 36(2): 168–177.

21. Gustaitis, J. 1988. Who Invented the Ice Cream Cone? *American History Illustrated* 23(4): 42–44.

22. Hamilton, S. 2003. The Economies and Conveniences of Modern-day Living: Frozen Foods and Mass Marketing, 1945–1965. *Business History Review* 77(1): 33–60.

23. Hawkins, R. 1995. The Baltimore Canning Industry and the Bahamian Pineapple Trade, 1865–1926. *Maryland Historian* 26(2): 1–22.

24. Heitmann, J. A. 1987. *The Modernization of the Louisiana Sugar Industry, 1830–1910.* Baton Rouge: Louisiana State University Press.

25. Hilliard, S. B. 1979. Site Characteristics and Spatial Stability of the Louisiana Sugarcane Industry. *Agricultural History* 53(1): 254–269.

26. Huang, C. J., et al. 1980. The Demand for Coffee in the United States 1963–1977. *Quarterly Review of Economics and Business* 20(2): 36–50.

27. Hunter, B. 2002. Rage for Grain: Flour Milling in the Mid-Atlantic, 1750–1815. Dissertation, University of Delaware.

28. Irwin, H. S. 1961. Some Early Chicago Butter Marketing Practices. *Agricultural History* 35(2): 82–84.

29. Jensen, J. M. 1982. Canning Comes to New Mexico: Women and the Agricultural Extension Service, 1914–1919. *New Mexico Historical Review* 57(4): 361–386.

30. ———. 1988. Butter Making and Economic Development in mid-Atlantic America from 1750 to 1850. *Signs* 13(4): 813–829.

31. Keuchel, E. F. 1976. The Polish American Migrant Worker: The New York Canning Industry, 1900–1935. *Polish American Studies* 33(2): 43–51.

32. Marks, E. 1977. The Story of Ice Cream. *Early American Life* 8(3): 36–37.

33. May, W. J. 1982. *The Great Western Sugarlands: History of the Great Western Sugar Company.* Dissertation, University of Colorado, Boulder.

34. Maclennan, C. A. 1995. Foundations of Sugar's Power: Early Maui Plantations, 1840–1860. *Hawaiian Journal of History* 29: 33–56.

35. Mepham, T. B. 1993. Humanizing Milk: The Formulation of Artificial Feeds for Infants, 1850–1910. *Medical History* 37(3): 225–249.

36. Moore, J. R. 2000. *Grist for the Mill: An Entrepreneurial History of Louisiana State Rice Milling Company, 1911–1965, River Brand Rice Milling Company, 1946–1965, and Riviana Foods, 1965–1999.* Lafayette: Center for Louisiana Studies.

37. Moyer, J. N. 2003. From Dairy to Doorstep: The Processing and Sale of New Hampshire Dairy Products, 1860s–1960s. *Historical New Hampshire* 58(3–4): 101–122.

38. Oakes, E. F. 1981. Baker's Chocolate: The Making of a Name. *Historical Journal of Massachusetts* 9(2): 29–40.

39. Patterson, G. 1997. Raising Cane and Refining Sugar: Florida's Crystals and the Fame of Fellsmere. *Florida Historical Quarterly* 75(4): 408–428.

40. Pendergrast, M. 1999. *Uncommon Grounds: The History of Coffee and How It Transformed Our World.* New York: Basic Books.

41. Persigehl, E. S. and York, J. D. 1979. Substantial Productivity Gains in the Fluid Milk Industry. *Monthly Labor Review* 102(7): 22–27.

42. Platt, C. V. and Meszaros, G. 1991. On Branch and Vine: Ohio's Fruits and Nuts. *Timeline* 8(4): 34–41.

43. Reidy, J. P. 1998. Mules and Machines and Men: Field Labor on Louisiana Sugar Plantations, 1887–1915. *Agricultural History* 72(2): 183–196.

44. Robbins, H. E. 2003. Jewish Bakery History: Los Angeles 1849–1926. *Western States Jewish History* 35(2): 122–143.

45. Roesti, R. M. 1965. An Economic Analysis of the Tuna Canning Industry. *Journal of the West* 4(4): 522–538.

46. Rosentreter, R. L. 1999. Cereal City. *Michigan History Magazine* 83(4): 6–13.

47. Schmalz, C. L. 1989. Sugar Beets in Cache Valley: An Amalgamation of Agriculture and Industry. *Utah Historical Quarterly* 57(4): 370–388.

48. Schmitz, M. 1979. The Transformation of the Southern Cane Sugar Sector, 1860–1930. *Agricultural History* 53(1): 270–285.

49. Sharer, G., et al. 1982. The Merchant-Millers: Baltimore's Flour Milling Industry, 1783–1860. *Agricultural History* 56(1): 167–171.

50. Spencer, J. J. 1984. To Make a Good Product Better: The Fairmont Creamery Company, 1884–1984. *Nebraska History* 65(3): 387–394.

51. Staible, D. R. 1990. Bakery Products. In *Manufacturing: A Historiographical and Bibliographical Guide,* ed. D. O. Whitten, 435–455. Westport, CT: Greenwood Press.

52. Volti, R. 1994. How We Got Frozen Food. *American Heritage of Invention & Technology* 9(4): 46–56.

53. Walsh, W. I. 1986. *The Rise and Decline of the Great Atlantic and Pacific Tea Company.* Secaucus, NJ: Lyle Stuart.

54. Webb, M. and Penner, K. P. 2000. *History of Food Irradiation* (MF-2426). Manhattan, KS: Kansas State University.

55. Weigley, E. S. 1999. Adventures in Good Eating: Duncan Hines of Kentucky. *Register of the Kentucky Historical Society* 97(1): 27–41.

56. Wilder, K. A. 1973. The Potato Starch Mill of Wilton, New Hampshire. *New England Galaxy* 14(4): 16–21.

57. Woloson, W. A. 2002. *Refined Tastes: Sugar, Confectionery, and Consumers in 19th Century America*. Baltimore, MD: Johns Hopkins University Press.

READINGS IN HAZARDOUS WASTE MANAGEMENT AND POLLUTION PREVENTION

58. Dugan, G. L., et al. 1972. Recycling System for Poultry Waste. *Journal of the Water Pollution Control Federation* 44(3): 432–440.

59. Gingerich, E. N., et al. 2003. Waste Management: The Poultry Industry. *Journal of the American Veterinary Medical Association* 223(1): 55.

60. Goldhor, S. H. and Koppernaes, J. D. n.d. *A Seafood Processor's Guide to Water Management*. Boston: New England Fisheries Development Association.

61. Lueck, R. H. and Pilcher, R. W. 1941. Canning Fruit Juices—Technical Aspects. *Industrial & Engineering Chemistry* 33(3): 2982–300.

62. Pennsylvania Department of Environmental Resources. 1994. *The Food Processing Residual Management Manual* (2500–BK-DER-1649). Harrisburg, PA: Pennsylvania Department of Environmental Resources.

63. U.S. Environmental Protection Agency. 1979. *Overview of Environmental Control Measures and Problems in the Food Processing Industries* (EPA/600/2–79/009). Washington, DC: Office of Research & Development.

64. ——. 1999. *Multimedia Environmental Compliance Guide for Food Processors* (EPA/305/B-99/005). Washington, DC: Office of Enforcement & Compliance Assurance.

65. Washington State Department of Ecology. 1994. *Pollution Prevention in Fruit & Vegetable Food Processing Industry* (94–051). Olympia, WA: Washington State Department of Ecology.

66. Willey, R. 2001. Fats, Oils, and Greases: The Minimization and Treatment of Wastewaters Generated from Oil Refining and Margarine Production. *Ecotoxicology & Environmental Safety* 50(2): 127–133.

67. Woodard, F. E., et al. 1972. Abatement of Pollution from a Poultry Processing Plant. *Journal of the Water Pollution Control Federation* 44(10): 1909–1915.

READINGS IN HEALTH AND SAFETY

68. Abramson, M. J., et al. 2001. Respiratory Disorders and Allergies in Tea Packers. *Occupational Medicine* 51: 259–265.

69. Campbell, J. A., et al. 1983. Cheese Worker's Hypersensitivity Pneumonitis. *American Review of Respiratory Disease* 127(4): 495–496.

70. Droste, J., et al. 2005. Occupational Exposure among Bakery Workers. *Journal of Occupational & Environmental Medicine* 47(5): 458–465.

71. el-Gazzar, F. E. and Marth, E. H. 1992. Salmonella, Salmonellas and Dairy Foods: A Review. *Journal of Dairy Science* 75(9): 2327–2343.

72. Fiocchi, A., et al. 2004. Clinical Tolerance of Processed Foods. *Annals of Allergy Asthma & Immunology* 93(Supp. 3): S38–S46.

73. Friedman, M. 2003. Nutritional Consequences of Food Processing. *Forum on Nutrition* 56: 350–352.

74. Holm S., et al. 2002. Cleaning Frequency and the Microbial Load in Ice Cream. *International Journal of Food Science & Nutrition* 53(4): 337–342.

75. Ikeda, K. 2002. New Seasonings. *Chemical Senses* 27(9): 847–849.

76. Jacobs, D. E. and Smith, M. S. 1988. Exposures to Carbon Dioxide in the Poultry Processing Industry. *AIHA Journal* 49(12): 624–629.

77. Jeebhay, M. F., et al. 2001. Occupational Seafood Allergy: A Review. *Occupational & Environmental Medicine* 58(9): 553–562.

78. Johnson, E. S., et al. 1997. Mortality from Cancer and Other Diseases in Poultry Slaughtering/Processing Plants. *International Journal of Epidemiology* 26(6): 1142–1150.

79. Larese, F., et al. 1998. Sensitization to Green Coffee Beans and Work Related Allergic Symptoms in Coffee Workers. *American Journal of Industrial Medicine* 34(6): 623–627.

80. Lehrer, S. B., et al. 1978. Extraction and Analysis of Coffee Bean Allergens. *Clinical Allergy* 8(3): 217–226.

81. McClean, D. and Pearce, N. 2004. Cancer among Meat Industry Workers. *Scandinavian Journal of Work & Environmental Health* 30(6): 425–437.

82. NIOSH. 2004. *Preventing Lung Disease in Workers Who Use or Make Flavorings* (Pub. 2004–165). Washington, DC: NIOSH.

83. Ropkins, K. and Beck, A. J. 2002. Controlling Organic Chemical Hazards in Food Manufacturing. *Nahrung* 46(4): 258–269.

84. Shapiro, A. and Mercier, C. 1994. Safe Food Manufacturing. *Science of the Total Environment* 143(1): 75–92.

85. Smith, T. 1999. Respiratory Symptoms, Sensitization and Dust Exposure in Small Bakeries. *Occupational Medicine* 49(8): 568–570.

86. Smith, T. A. 2004. Incidence of Occupational Skin Conditions in a Food Manufacturing Company. *Occupational Medicine* 54(4): 227–230.

87. Smith, T. A. and Lumley, K. P. 1996. Work-Related Asthma in a Population Exposed to Grain, Flour and Other Ingredient Dusts. *Occupational Medicine* 46(1): 37–40.

88. Smith, T. A., et al. 2000. Respiratory Symptoms and Wheat Flour Exposure: A Study of Flour Millers. *Occupational Medicine* 50(1): 25–29.

89. Thaler, A. M. 1999. The U.S. Perspective towards Poultry Slaughter. *Poultry Science* 78(2): 298–301.

90. Tovey, F. I. and Hobsley, M. 2004. Milling of Wheat, Maize and Rice: Effects on Fiber and Lipid Content and Health. *World Journal of Gastroenterology* 10(12): 1695–1696.

91. Vanhanen, M., et al. 1996. Enzyme Exposure and Enzyme Sensitization in the Baking Industry. *Occupational & Environmental Medicine* 53(10): 670–676.

92. Waldroup, A. L. 1993. Summary of Work to Control Pathogens in Poultry Processing. *Poultry Science* 72(6): 1177–1179.

93. Wax, P. M. 1995. Tortilla Thallotoxicosis. *Journal of Toxicology & Clinical Toxicology* 33(3): 265.

94. Zuskin, E., et al. 1993. Respiratory Symptoms and Ventilatory Capacity in Workers in a Vegetable Pickling and Mustard Production Facility. *International Archives of Occupational & Environmental Health* 64(6): 457–461.

95. ——. 1994. Immunologic Findings in Confectionary Workers. *Annals of Allergy* 73(6): 521–526.

312: Beverage and Tobacco Product Manufacturing

READINGS IN HISTORY

96. Anderson, C. 1977. The Great American Soft Drink. *Mankind* 5(11): 8–13.

97. Angus, B. J. and Harris, R. 1993. The Bottling Industries of Early El Paso. *Password* 38(3): 143–151.

98. Brady, R. 1979. Vintage Wineries. *Historic Preservation* 31(4): 2–10.

99. Burns, M. R. 1983. Economies of Scale in Tobacco Manufacture, 1897–1910. *Journal of Economic History* 43(2): 461–474.

100. Clay, K. and Troesken, W. 2002. Strategic Behavior in Whiskey Distilling, 1887–1895. *Journal of Economic History* 62(4): 999–1023.

101. Damstra, C. 2001. The Great Grape State. *Michigan History Magazine* 85(5): 28–37.

102. Demler, F. R. 1981. The Nature of Tin Substitution in the Beverage Container Industries. Dissertation, Pennsylvania State University.

103. Dick, C. E. 1981. A Geographical Analysis of the Development of the Brewing Industry of Minnesota. Dissertation, University of Minnesota.

104. Downward, W. L. 1973. *The Cincinnati Brewing Industry: A Social and Economic History*. Athens: University of Ohio Press.

105. ——. 1980. *Dictionary of the History of the American Brewing & Distilling Industries*. Westport, CT: Greenwood Press.

106. Gaston, K. B. 1984. Robertson County Distilleries, 1796–1909. *Tennessee Historical Quarterly* 43(1): 49–67.

107. ——. 1998. George Dickel Tennessee Sour Mash Whiskey: The Story Behind the Label. *Tennessee Historical Quarterly* 57(2): 150–167.

108. Geraci, V. W. 2000. The Family Wine-farm: Vintibusiness Style. *Agricultural History* 74(2): 419–432.

109. Hull-Walski, D. A. and Walski, F. L. 1994. There's Trouble a' Brewin': The Brewing and Bottling Industries at Harpers's Ferry, West Virginia. *Historical Archeology* 28(4): 106–121.

110. Irwin, N. L. 1992. Bottling Gold: Chattanooga's Coca-Cola Fortunes. *Tennessee Historical Quarterly* 51(4): 223–237.

111. Kerr, K. A. 1990. Distilled Spirits. In *Manufacturing: A Historiographical and Bibliographical Guide,* ed. D. O. Whitten, 56–62. Westport, CT: Greenwood Press.

112. Knapp, V. 1974. The Natural Ice Industry of Philadelphia in the 19th Century. *Pennsylvania History* 41(4): 413–421.

113. Lochart, B. 1999. From "Chasers" to the Family Trade: Some Early El Paso Soda Bottlers. *Password* 44(4): 173–180.

114. McGahan, A. M. 1991. The Emergence of the National Brewing Oligopoly: Competition in the American Market, 1933–1958. *Business History Review* 65(2): 229–284.

115. Peters, G. L. 1984. Trends in California Viticulture. *Geographical Review* 74(4): 455–467.

116. Pezzoni, J. D. 2001. The Architecture of Tobacco Manufacturing in 19th Century Virginia. *Pioneer America Society Transactions* 24: 69–81.

117. Purcell, A. D. 1998. Bourbon to Bullets: Louisville's Distilling Industry during World War II. *Register of the Kentucky Historical Society* 96(1): 61–87.

118. Roberts, C. 1984. Winemaking in the Mother Lode. *Westways* 76(9): 38–40.

119. Roberts, W. C. and Knapp, R. F. 1992. Paving the Way for the Tobacco Trust: From Hand Rolling to Mechanized Cigarette Production by W. Duke, Sons & Company. *North Carolina Historical Review* 69(3): 256–281.

120. Ronnenberg, H. 1993. *Beer and Brewing in the Inland Northwest, 1850–1950*. Moscow: University of Idaho Press.

121. Ronnenberg, H. W. 1990. History of the Brewing Industry in Idaho, 1862–1960. Dissertation, University of Idaho.

122. Schnell, S. M. and Reese, J. F. 2003. Microbreweries as Tools of Local Identity. *Journal of Cultural Geography* 21(1): 45–70.

123. Shaner, R. H. 1963. Distillation and Distilleries among the Dutch. *Pennsylvania Folklife* 13(3): 39–42.

124. Sharpe, C. C. 1997. The Spearman Brewing Company. *Pensacola History Illustrated* 5(2): 1–4.

125. Slack, M. 2000. Local and Regional Breweries in America's Brewing Industry, 1865–1920. *Business History Review* 74(3): 435–463.

126. Stack, M. H. 1999. Liquid Bread: An Examination of the American Brewing Industry, 1865–1940. Dissertation, University of Notre Dame.

127. Taplin, I. M. 1992. Rising from the Ashes: The Deskilling Debate and Tobacco Manufacturing. *Social Science Journal* 29(1): 87–106.

READINGS IN HAZARDOUS WASTE MANAGEMENT AND POLLUTION PREVENTION

128. Boruff, C. S. 1939. Waste Problems in the Fermentation Industry. *Industrial & Engineering Chemistry* 31(11): 1335–1337.

129. Hoover, C. R. and Burr, F. K. 1936. Distillery Wastes: Chemical and Filtration Studies. *Industrial & Engineering Chemistry*, pp. 38–42.

130. Lammers, F. J. 1934. Water Purification in the Modern Brewery and Distillery. *Industrial & Engineering Chemistry* 26(11): 1133–1138.

131. Lange, C. R. 1993. Fermentation Industry. *Water Environment Research* 65(4): 400–402.

132. Paulette, R. G., et al. 1970. A Pollution Abatement Program for Distillery Wastes. *Journal of the Water Pollution Control Federation* 42(7): 1387–1394.

133. Rodrigues, A. C., et al. 2001. Post-Treatment of Brewery Wastewater Using a Sequencing Batch Reactor. *Water Environment Research* 73(1): 45–51.

134. Sefcova, H. 1999. Microbiological Control of the Water Bottling Process. *Central European Journal of Public Health* 7(4): 182–184.

135. Tofflemire, T. J. 1972. Survey of Methods of Treating Wine and Grape Wastewater. *American Journal of Enology and Viticulture* 23(4): 165–172.

136. Vaughn, R. H. and Marsh, G. L. 1953. Disposal of California Winery Wastes. *Industrial & Engineering Chemistry* 45(12): 2686–2688.

READINGS IN HEALTH AND SAFETY

137. Dodson, R. F., et al. 2002. Asbestos Burden in Two Cases of Mesothelioma Where the Work History Included Manufacturing of Cigarette Filters. *Journal of Toxicology & Environmental Health A* 65(16): 1109–1120.

138. Ghosh, S. K., et al. 1986. Occupational Health Problems among Workers Handling Virginia Tobacco. *International Archives of Occupational & Environmental Health* 58(1): 47–52.

139. Glindmeyer, H. W., et al. 1991. Exposure Related Declines in the Lung Function of Cotton Textile Workers. *American Review of Respiratory Disease* 144(Part 1): 675–683.

140. Godnic-Cvar, J., et al. 1999. Respiratory and Immunological Findings in Brewery Workers. *American Journal of Industrial Medicine* 35(1): 68–75.

141. Hoffmann, D. and Hoffmann, I. 1997. The Changing Cigarette, 1950–1995. *Journal of Toxicology & Environmental Health* 50: 307–364.

142. Huncharek, M. 1994. Pleural Mesothelioma in a Cigarette Factory Worker. *Scandinavian Journal of Work & Environmental Health* 20(2): 146–147.

143. Kalogeromitros, D., et al. 2004. Asymptomatic Sensitization to Grapes in a Sample of Workers in the Wine Industry. *Occupational & Environmental Medicine* 61(8): 709–711.

144. Kaufmann, A. 1998. Lead in Wine. *Food Additives & Contamination* 15(4): 437–445.

145. Love, R. G., et al. 1988. Respiratory and Allergic Symptoms in Wool Textile Workers. *British Journal of Industrial Medicine* 45(11): 727–741.

146. Mondino, B. J., et al. 1978. Ocular Injuries from Exploding Beverage Bottles. *Archives of Ophthalmology* 96(11): 2040–2041.

147. Personick, M. E. and Harthun, L. A. 1992. Profiles in Safety and Health: The Soft Drink Industry. *Monthly Labor Review* 115(4): 12–17.

148. Sorenson, G., et al. 1997. Worksite Characteristics and Changes in Worksite Tobacco-controlled Initiatives. *Journal of Occupational Medicine* 39(6): 520–526.

149. Talcott, J. A., et al. 1989. Asbestos-associated Diseases in a Cohort of Cigarette-filter Workers. *NEJM* 321(18): 1220–1223.

150. Trape-Cardoso, M., et al. 2005. Cotinine Levels and Green Tobacco Sickness among Shade-tobacco Workers. *Journal of Agromedicine* 10(2): 27–37.

151. Uchikoshi, S., et al. 1981. Nasal Allergy in Spinning Mill Workers. *Tokai Journal of Experimental & Clinical Medicine* 6(4): 363–371.

152. WorkCover, N. W. S. 1999. *Wine Industry Code of Practice for Workplace Health & Safety* (No. 129). http://www.workcover.nsw.gov.au/Documents/Publications/LawandPolicy/CodesOfPractice/wine_industry_code_of_practice_0129.pdf (accessed 17 October 2008).

153. Zuskin, E., et al. 2004. Immunological and Respiratory Changes in Tobacco Workers. *American Journal of Industrial Medicine* 45(1): 76–83.

313: Textile Mills
3131: Fiber, Yarn, and Thread Mills
3132: Fabric Mills
3133: Textile and Fabric Finishing and Fabric Coating Mills

READINGS IN HISTORY

154. Andrews, M. G. 1988. *The Men and the Mills: A History of the Southern Textile Industry*. Macon, GA: Mercer University Press.

155. Anonymous. 1963. A List of Locations of American Textile Mills, 1808–1818. *Textile History Review* 4(1): 46–51.

156. Jeremy, D. J. 1971. British and American Yarn Count Systems: An Historical Analysis. *Business History Review* 45(3): 336–368.

157. Lomax, A. L. 1961. The Portland Woolen Mills, Inc. *Oregon Historical Quarterly* 62(2): 164–179.

158. Yearous, J. 1998. Stitches in Time: The Development of Sewing Thread in the 19th Century and Beyond. *Uncoverings* 19: 155–178.

159. Zeisel, R. N. 1973. Modernization and Manpower in Textile Mills. *Monthly Labor Review* 96(6): 18–25.

READINGS IN HAZARDOUS WASTE MANAGEMENT AND POLLUTION PREVENTION

160. Bajaj, P. 2001. Finishing of Textile Materials. *Journal of Applied Polymer Science* 83(3): 631–659.

161. Bal, A. S. 1999. Wastewater Management for Textile Industry: An Overview. *Indian Journal of Environmental Health* 41(4): 264–290.

162. Cordova, R. E. V., et al. 2001. Toxicity-based Criteria for the Evaluation of Textile Wastewater Treatment Efficiency. *Environmental Toxicology & Chemistry* 20(4): 839–845.

163. Newlin, K. D. 1971. Economic Feasibility of Treating Textile Wastes in Municipal Systems. *Journal of the Water Pollution Control Federation* 43(11): 2195–2199.

164. Smith, B. 1990. *Identification and Reduction of Toxic Pollutants in Textile Mill Effluents*. Raleigh, NC: Office of Waste Reduction, Department of Environment, Health, and Natural Resources.

165. U.S. Environmental Protection Agency. 1997. *Profile of the Textile Industry*. Rockville, MD: ABS Group.

166. ——. 2002. *RCRA in Focus: Textile Manufacturing* (EPA/530/K-02/028). Washington, DC: Office of Solid Waste & Emergency Response.

167. Vandevivere, P. C., et al. 1998. Treatment and Reuse of Wastewater from Textile Wet-processing Industry: Review of Emerging Technologies. *Journal of Chemical Technology & Biotechnology* 72(4): 289–302.

READINGS IN HEALTH AND SAFETY

168. Abdel-Kader, H. M., et al. 1987. Long-Term Cotton Dust Exposure in the Textile Industry. *AIHA Journal* 48(6): 545–550.

169. Christiani, D.C., et al. 2001. Longitudinal Changes in Pulmonary Function and Respiratory Symptoms in Cotton Textile Workers. A 15-Year Follow-up Study. *American Journal of Respiratory & Critical Care Medicine* 163(4): 847–853.

170. Drews, M. J. and Hatcher, J. D. 1983. A Chemical and Engineering Analysis of Dust in Textile Mills. *AIHA Journal* 44(12): 903–910.

171. Gibbs, G. W. 1994. The Assessment of Exposure in Terms of Fibers. *Annals of Occupational Hygiene* 38(4): 477–487.

172. IARC. 1990. *IARC Monographs on the Evaluation of Carcinogenic Risks to Humans: Some Flame Retardants and Textile Chemicals, and Exposures in the Textile Manufacturing Industry* (Monograph No. 48). Lyon, France.

173. Mills, R. O. 1969. Noise Reduction in a Textile Weaving Mill. *AIHA Journal* 30(1): 71–76.

174. Sorainen, E. 1988. Vibration in Textile Mills. *AIHA Journal* 49(12): A775–A778.

175. Zuskin, E., et al. 1995. Respiratory Symptoms and Lung Function in Wool Textile Workers. *American Journal of Industrial Medicine* 27(6): 845–857.

314: Textile Product Mills
3141: Textile Furnishings Mills
3149: Other Textile Product Mills

READINGS IN HISTORY

176. Evans, S. 2001a. Prison Made Binder Twine. *North Dakota History* 68(1): 20–36.

177. ——. 2001b. From Kanasin to Kansas: Mexican Sisal, Binder Twine, and the State Penitentiary Twine Factory, 1890–1940. *Kansas History* 24(4): 276–299.

178. Friedman, T. J. 2001. Communities in Competition: Capital Migration and Plant Relocation in the U.S. Carpet Industry, 1929–1975. Dissertation, Columbia University.

179. Merry, B. A. and Martinez, B. 1991. Rope. *American Heritage of Invention & Technology* 7(2): 38–45.

180. Patton, R. L. and Parker, D. B. 1999. *Carpet Capital: The Rise of a New South Industry.* Athens: University of Georgia Press.

READINGS IN HAZARDOUS WASTE MANAGEMENT AND POLLUTION PREVENTION

181. Defrawy, N. M. H. 2002. Water Management in the Textile Industry: Technical and Economic Aspects. *International Journal of Environmental Studies* 59(5): 573–587.

182. Eisemann, K. and Young, W. M. 1960. Study of a Textile Mill with the Aid of Linear Programming. *Management Technology* 1(1): 52–63.

183. Leonas, K. K. and Leonas, M. L. 1994. Textile Process Wastewater Permits: An Update and Strategies. *American Dyestuff Reporter* (March): 26–34.

184. Netzer, A. and Beszedits, S. 1975. Pollution from Textile Mills: Problems and Solutions. *Water Pollution Control* 113 (5): 17–18.

185. Sarkar, A. 1998. Implementation of ISO 9000 in a Textile Mill. *Total Quality Management* 9(1): 123–131.

186. Smith, B. 1986. *Identification and Reduction of Pollution Sources in Textile Wet Processing*. Raleigh, NC: North Carolina State University, Department of Textile Chemistry, School of Textiles.

187. Tu, Y. and Yeung, E. H. H. 2005. Integrated Maintenance Management System in a Textile Company. *International Journal of Advanced Manufacturing Technology* 13(6): 453–461.

188. Wynne, G., et al. 2000. *Cleaner Production in the Textile Industry—Lessons from the Danish Experience.* Natal, South Africa: University of Natal, Pollution Research Group.

READINGS IN HEALTH AND SAFETY

189. Abel-Kader, H. M., et al. 1987. Long-Term Cotton Dust Exposure in the Textile Industry. *AIHA Journal* 48(6): 545–550.

190. Ferguson, R. G., et al. 1977. Quieting a Wire Strander for the Tire Cord Industry. *AIHA Journal* 38(7): 321–326.

191. IARC. 1990. *IARC Monographs on the Evaluation of Carcinogenic Risks to Humans: Some Flame Retardants and Textile Chemicals, and Exposures in the Textile Manufacturing Industry* (Monograph No. 48). Lyon, France.

192. Malo, J. L., et al. 1990. Prevalence of Occupational Asthma and Immunologic Sensitization to Guar Gum among Employees at a Carpet Manufacturing Plant. *Journal of Allergy & Clinical Immunology* 86(4) (Pt. 1): 562–569.

193. Quinn, M. M., et al. 1987. An Asbestos Hazard in the Reprocessed Textile Industry. *American Journal of Industrial Medicine* 11(3): 255–266.

315: Apparel Manufacturing
3151: Apparel Knitting Mills
3152: Cut and Sew Apparel Manufacturing

READINGS IN HISTORY

194. Anderson, M. G. 1986. Munsingwear: An Underwear for America. *Minnesota History* 50(4): 152–161.

195. Bates, C. 2000. Women's Hats and the Millinery Trade, 1840–1940: An Annotated Bibliography. *Dress* 27: 49–58.

196. Blicksilver, J. 1990. Apparel and Other Textile Products. In *Manufacturing: A Historiographical and Bibliographical Guide,* ed. D. O. Whitten, 89–114. Westport, CT: Greenwood Press.

197. Candee, R. M. 1997. The Shaker Sock: A New Hampshire Contribution to 19th Century Machine Knitting. *Historical New Hampshire* 52(3–4): 60–77.

198. ———. 2000. British Framework Knitters in New England: Technology Transfer and Machine Knitting in America, 1820–1900. *Textile History* 31(1): 27–53.

199. ———. 2002. Old Portsmouth . . . Home of the Stocking Business: A Forgotten 19th Century Industry and Its Inventors. *Historical New Hampshire* 57(3–4): 85–107.

200. Chance, L. 1983. The Hats on Early American's Heads. *Early American Life* 14(6): 58–60, 70.

201. Daily, C. 1982. A Woman's Concern: Millinery in Central Iowa, 1870–1880. *Journal of the West* 21(2): 26–32.

202. DeMoss, D. D. 1981. The History of Apparel Manufacturing in Texas, 1897–1981. Dissertation, Texas Christian University.

203. Dewhurst, K. C. 1988. Pleiku Jackets, Tour Jackets, and Working Jackets: The Letter Sweaters of War. *Journal of American Folklore* 101(399): 48–52.

204. Farrell-Beck, J. 1992. Use of Patented Sewing Devices in a Sample of Manufactured Garments Dated ca. 1860–1900. *Clothing & Textile Research Journal* 10(3): 3–11.

205. Fields, J. S. 1998. The Production of Glamour: A Social History of Intimate Apparel, 1909–1952. Dissertation, University of Southern California.

206. Fox, J. S. 1982. The Location Dynamics of the United States Knitting Industry. Dissertation, University of Tennessee.

207. Gamber, W. 1997. *The Female Economy: The Millinery and Dressmaking Trades, 1860–1930*. Urbana: University of Illinois Press.

208. Haas, W. A. 1971. Levi's Pants Factory. *Historic Preservation* 23(4): 14–17.

209. Hardin, A. L. 1990. Industry Structure and the Marketing of Synthetic Fibers. *Business & Economic History* 19: 213–222.

210. Huun, K. M-A. 1995. Men's Neckties of the 20th Century: A History, Semiotic Analysis and Data Classification System. Dissertation, Florida State University.

211. Josephy, A. M. 1985. Those Pants That Levi Gave Us. *American West* 22(4): 30–37.

212. LaChapelle, P. 2001. All That Glitters: Country Music, Taste, and the Politics of the Rhinestone Nudie Suit. *Dress* 28: 3–12.

213. Lintner, M. D. L. 1979. The Height of Fashion: The Construction of Ladies Fashion Headwear, 1830–1914. Dissertation, University of Michigan.

214. Lowe, E. D. and Lowe, J. W. G. 1990. Velocity of the Fashion Process in Women's Formal Evening Dress, 1789–1980. *Clothing & Textiles Research Journal* 9(1): 50–58.

215. Phillips, J. and Phillips, P. 1993. History from Below: Women's Underwear and the Rise of Women's Sports. *Journal of Popular Culture* 27(2): 129–148.

216. Pivar, D. J. 1964. The Hosiery Workers and the Philadelphia Third Party Impulse, 1929–1935. *Labor History* 5(1): 18–28.

217. Rosen, E. I. 2002. *Making Sweatshops: The Globalization of U.S. Apparel Industry.* Berkeley: University of California Press.

218. Shine, C. R. 1987. Hunting Shirts and Silk Stockings: Clothing Early Cincinnati. *Queen City Heritage* 45(3): 23–48.

219. Turbin, C. 2002. Fashioning the American Man: The Arrow Collar Man, 1907–1931. *Gender & History* 14(3): 470–491.

220. Wehrle, L. 1989. The Development of the American Knitting Machine Needle to 1900. *Ars Textrina* 12: 133–155.

221. Weiler, S. N. 1979. Walkout: The Chicago Men's Garment Worker's Strike, 1910–1911. *Chicago History* 8(4): 238–249.

222. Williams, D. C. 1997. Those Pants of Levi's: A Great Western Enterprise. *Journal of the West* 36(1): 77–81.

223. Wong, Y. Y. 1992. Adapting Organizational Strategies to Evolving External Environment: A Multiple-Case Study of U.S. Apparel Manufactures. Dissertation, University of Nebraska, Lincoln.

224. Workman, N. V. 1996. From Victorian to Victoria's Secret: The Foundations of Modern Erotic Wear. *Journal of Popular Culture* 30(2): 61–73.

225. Wright, A. C. 1995. Strategy and Structure in the Textile Industry: Spencer Love and Burlington Mills, 1923–1962. *Business History Review* 69(1): 42–79.

226. Zakim, M. 1998. Customizing the Industrial Revolution: The Reinvention of Tailoring in the 19th Century. *Winterthur Portfolio* 33(1): 41–58.

READINGS IN HAZARDOUS WASTE MANAGEMENT AND POLLUTION PREVENTION

227. Beghin, J. and Potier, M. 1997. Effects of Trade Liberalization on the Environment in the Manufacturing Sector. *World Economy* 20(4): 435–456.

228. Edwards, F. G., et al. 2006. Electrolytic Treatment of an Industrial Wastewater from a Hosiery Plant. *Water Environment Research* 78(4): 435–441.

229. Guner, M. and Yucel, O. 2005. Environmental Protection and Waste Management in Textile and Apparel Sectors. *Journal of Applied Sciences* 5(10): 1843–1849.

230. Jones, R. M. 2002. *The Apparel Industry*. Oxford: Blackwell Science.

231. Nimon, W. and Beghin, J. 1999. Are Eco-labels Valuable? Evidence from the Apparel Industry. *American Journal of Agricultural Economics* 81(4): 801–811.

READINGS IN HEALTH AND SAFETY

232. Dignan, M., et al. 1996. Cumulative Trauma Disorders among Apparel Manufacturing Employees in the Southeast United States. *Southern Medical Journal* 89(11): 1057–1062.

233. Hatch, K. L., et al. 1986. Textile Chemical Finish Dermatitis. *Contact Dermatitis* 14(1): 1–13.

234. Malker, H. R., et al. 1984. Kidney Cancer among Leather Workers. *Lancet* 1(8367): 56.

235. Savitz, D. A., et al. 1996. Male and Female Employment in the Textile Industry in Relation to Miscarriage and Preterm Delivery. *American Journal of Industrial Medicine* 30(3): 307–316.

236. Silberman, D. E. and Sorrell, A. H. 1959. Allergy in Fur Workers with Special Reference to Paraphenylenediamine. *Journal of Allergy* 30(1): 11–18.

237. Sweeney, M. H., et al. 1985. Mortality among Retired Fur Workers. *Scandinavian Journal of Work & Environmental Health* 11(4): 257–264.

316: Leather and Allied Product Manufacturing
3161: Leather and Hide Tanning and Finishing
3162: Footwear Manufacturing
3169: Other Leather and Allied Product Manufacturing

READINGS IN HISTORY

238. Duke, J. and Usher, L. 1989. Multifactor Productivity Slips in the Non-rubber Footwear Industry. *Monthly Labor Review* 112(4): 32–38.

239. Ellsworth, L. F. 1965. The Delaware Leather Industry in the Mid-19th Century. *Delaware History* 11(4): 261–281.

240. Grobbel, W. B. 1997. Tanning and Tanneries: An Introduction to Archaeology of the Leather Industry. *North American Archaeologist* 18(2): 121–147.

241. Gurian, B. 2002. The Impact of the Sneaker and Sneaker Industry on Modern Society: A Review of the Transformative Powers of an Icon. Dissertation, St. John's University.

242. Koffler, S. 2003. American Tourister Luggage. *Rhode Island Jewish Historical Notes* 14(1): 72–79.

243. McCormick, M. 1988. Handmade Trunks. *Annal of Wyoming* 60(1): 17–21.

244. Miranda, J. A. 2004. American Machinery and European Footwear: Technology Transfer and International Trade, 1860–1939. *Business History* 46(2): 195–218.

245. Rotenstein, D. S. 1997a. From Farm to Factory: Craft Dynasties and Leather Tanning in 19th Century Pennsylvania. Dissertation, University of Pennsylvania.

246. Rotenstein, D. S. 1997b. Leather Bound: 19th Century Leather Tanners in Allegheny City. *Pittsburg History* 80(1): 32–48.

247. Sorenson, O. and Audia, P. G. 2000. The Social Structure of Entrepreneurial Activity: Geographic Concentration of Footwear Production in the United States, 1940–1989. *American Journal of Sociology* 106(2): 424–461.

248. Stiles, D. 1998. Martin Butler, Masculinity, and the North American Sole Leather Tanning Industry, 1871–1889. *Labour* 42: 85–114.

249. Wiles, J. L. 1990. Leather and Leather Products. In *Manufacturing: A Historiographical and Bibliographical Guide,* ed. D. O. Whitten, 213–235. Westport, CT: Greenwood Press.

READINGS IN HAZARDOUS WASTE MANAGEMENT AND POLLUTION PREVENTION

250. Beghin, J. and Potier, M. 1997. Effects of Trade Liberalization on the Environment in the Manufacturing Sector. *World Economy* 20(4): 435–456.

251. Environment Agency. 2001. *Integrated Pollution Prevention and Control: Guidance for the Tanneries Sector* (IPPC S6.08). http://www.sepa.org.uk/ppc/uktech/index.htm (accessed 17 October 2008).

252. Simenova, L. S. and Dalev, P. G. 1996. Utilization of Leather Industry Waste. *Waste Management* 16(8): 765–770.

253. Tesarik, I. and Vostrcil, J. 1972. Measures Taken against Water Pollution in Tanneries and Leather Finishing Plants. *Pure Applied Chemistry* 29(1): 365–379.

254. Thanikaivelan P., et al. 2005. Recent Trends in Leather Making: Processes, Problems and Pathways. *Critical Reviews in Environmental Science & Technology* 35(1): 37–79.

255. U.S. Environmental Protection Agency. 2000. *RCRA in Focus: Leather Manufacturing* (EPA/530/K-00/002). Washington, DC: Office of Solid Waste & Emergency Response.

READINGS IN HEALTH AND SAFETY

256. Battista, G., et al. 1995. Nasal Cancer in Leather Workers: An Occupational Disease. *Journal of Cancer Research & Clinical Oncology* 121(1): 1–6.

257. IARC. 1981. *IARC Monographs on the Evaluation of Carcinogenic Risks to Humans: Wood, Leather and Some Associated Industries* (Monograph No. 25). Lyon, France.

258. Mancuso, G., et al. 1996. Occupational Dermatitis in Shoemakers. *Contact Dermatitis* 34(1): 17–22.

259. Montanaro, F., et al. 1997. Mortality in a Cohort of Tannery Workers. *Occupational and Environmental Medicine* 54(8): 588–591.

260. Pippard, E. C., et al. 1985. Mortality of Tanners. *British Journal of Industrial Medicine* 42(4): 285–287.

261. Rowland, A. 1980. Tanning Leather, Tanning Hides: Health and Safety Struggle in a Leather Factory. *Radical America* 14(6): 23–37.

262. Stern, F. B. 2003. Mortality among Chrome Leather Tannery Workers: An Update. *American Journal of Industrial Medicine* 44(2): 197–206.

321: Wood Product Manufacturing
3211: Sawmills and Wood Preservation
3212: Veneer, Plywood, and Engineered Wood Product Manufacturing
3219: Other Wood Product Manufacturing

READINGS IN HISTORY

263. Ball, L. D. and Clements, W. M. 1992. The Very Last Timber: Harvesting Chemical Wood in Northeastern Arkansas. *Forest & Conservation History* 36(2): 73–81.

264. Barillas, W. 1989. Michigan's Pioneers and the Destruction of the Hardwood Forest. *Michigan Historical Review* 15(2): 1–22.

265. Bergmark, J. H. 1981. The Trailer Home: Transitional Shelter and Shelter in Transition. *Kentucky Folklore Record* 27(3–4): 97–102.

266. Callahan, J. C. 1990. *The Fine Hardwood Veneer Industry in the United States, 1838–1990.* Lake Ann, MI: National Woodlands.

267. Clark, D. L. 1986. The Indiana Hardwood Industry. Dissertation, Purdue University.

268. ——. 1987. *Entrepreneurs in Hardwood.* Indianapolis: Indiana Hardwood Lumberman's Association.

269. Craig, B. and Pencavel, J. 1992. The Behavior of Worker Cooperatives: The Plywood Companies of the Pacific Northwest. *American Economic Review* 82(5): 1083–1105.

270. Darwin, R. F. 1991. Measuring the Effects of Input-Reducing Technologies: Softwood Lumber Production in the Pacific Northwest, 1950–1985. Dissertation, University of Missouri, Columbia.

271. D'Olive, A. L. 1977. Reminiscences of Ten Mile: A South Mississippi Sawmill Town. *Journal of Mississippi History* 39(2): 173–184.

272. Downs, A. C. 1976. Reeve & Brothers, New Jersey Saw Millers and Ship Builders 1821–1859. *American Neptune*36(4): 251–256.

273. Eisterhold, J. A. 1973. Mobile: Lumber Center of the Gulf Coast. *Alabama Review* 26(2): 83–104.

274. Finges, J. R. 1972. Seattle's First Sawmill, 1853–1869: A Study of Frontier Enterprise. *Forest History* 15(4): 24–31.

275. French, R. M. and Hadden, J. K. 1965. An Analysis of the Distribution and Characteristics of Mobile Homes in America. *Land Economics* 41(2): 131–139.

276. Gray, E. 1986. *Roughing It on the Little Deschutes River, 1934–1944*. Eugene, OR: Edward Gray.

277. Hart, J. F. and Morgan, J. T. 1995. Mobile Homes. *Journal of Cultural Geography* 15(2): 35–53.

278. Hustar, D. A. 1975. Ohio Water-Powered Sawmills. *Ohio History* 84(1–2): 5–56.

279. Hyland, S. E. 1977. Mid-West Mobile Home Parks: An Emerging Residential Settlement Alternative. Dissertation, University of Illinois, Urbana-Champaign.

280. Johnson, M. C. 1997. Sawmills and Mill Towns of the Lanana Bayou, Nacogdoches County, Texas 1834 to 1910. *East Texas Historical Journal* 35(2): 3–11.

281. Knipping, M. 2002. Cutting Edge History: Reconstructing the Herrling Sawmill. *Wisconsin Magazine of History* 85(4): 20–31.

282. Linde, M. 1984. *Sawmills of the Black Hills*. Rapid City, SD: Sundstrom.

283. MacTavish, K. and Salamon, S. 2001. Mobile Home Park on the Prairie: A New Rural Community Form. *Rural Sociology* 66(4): 487–506.

284. Maxwell, R. S. 1982. The First Big Mill: The Beginnings of Commercial Lumbering in Texas. *Southwestern Historical Quarterly* 86(1): 1–30.

285. Miller, H. L. 1995. *Lumbering in Early 20th Century Michigan*. Lewiston, MI: Walnut Hill.

286. Namorato, M. V. 1990. Lumber and Wood Products. In *Manufacturing: A Historiographical and Bibliographical Guide,* ed. D. O. Whitten, 118–133. Westport, CT: Greenwood Press.

287. Oliver, E. S. 1978. Sawmilling on Gray's Harbor in the Twenties: A Personal Reminiscence. *Pacific Northwest Quarterly* 69(1): 1–18.

288. Richards, D. B. 1964. Prospects for a Southern Pine Plywood Industry in Alabama. *Journal of the Alabama Academy of Science* 35(4): 149–152.

289. Silver, J. W. 1957. The Hardwood Producer Comes of Age. *Journal of Southern History* 23(4): 427–453.

290. Sitton, T. and Conrad, J. H. 1998. *Nameless Towns: Texas Sawmill Communities, 1880–1942.* Austin: University of Texas Press.

291. Swiatosz, S. 1985. A Technical History of Late 19th Century Windows in the United States. *APT Bulletin* 17(1): 31–37.

292. Tamura, L. 1993. Railroads, Stumps and Sawmills: Japanese Settlers of the Hood River Valley. *Oregon Historical Quarterly* 94(4): 368–398.

293. Thornburg, D. A. 1991. *Galloping Bungalows: The Rise and Demise of the American House Trailer*. Hamden, CT: Archon Books.

294. vanWormer, S. 1983. The Wetherbee Planing Mill: A Case History of the 1880s Boom and Bust. *Journal of San Diego History* 29(1): 20–27.

295. Veigle, J. and Brand, H. 1982. Millwork Industry Shows Slow Growth in Productivity. *Monthly Labor Review* 105(9): 21–26.

296. Walker, D. R. 1996. Harvesting the Forest: Henry Jacob Lutcher, G. Bedell Moore and the Advent of Commercial Lumbering in Texas. *Journal of the West* 35(3): 10–17.

297. Wallis, A. D. 1991. *Wheel Estate: The Rise and Decline of Mobile Homes*. New York: Oxford University Press.

298. Wyckoff, W. 1986. Frontier Milling in Western New York. *Geographical Review* 76(1): 73–93.

READINGS IN HAZARDOUS WASTE MANAGEMENT AND POLLUTION PREVENTION

299. Beghin, J. and Potier, M. 1997. Effects of Trade Liberalization on the Environment in the Manufacturing Sector. *World Economy* 20(4): 435–456.

300. Grosse, D. W., et al. 2000. The Treatment of Contaminated Water at Remedial Wood Preserving Sites. *Remediation* 10(3): 111–128.

301. Hingston, J. A., et al. 2001. Leaching of Chromated Copper Arsenate Wood Preservatives: A Review. *Environmental Pollution* 111(1): 53–66.

302. Kansas Department of Health & Environment. 1999. *Pollution Prevention for the Wood Manufacturing & Finishing Industry*. http://64.233.169.104/search?q=cache:_QbKPEU2NQMJ:www.p2pays.org/ref/15/14845.pdf+%22Pollution+Prevention+for+the+Wood+Manufacturing+%26+Finishing+Industry%22&hl=en&ct=clnk&cd=1&gl=us (accessed 17 October 2008).

303. McKeever D. B. 1998. Wood Residual Quantities in the United States. *Biocycle* 39(1): 65–68.

304. Skog, K. E., et al. 1995. Wood Products Technology Trends: Changing the Face of Forestry. *Journal of Forestry* 93(12): 30–34.

305. U.S. Environmental Protection Agency. 1988. *Estimating Chemical Releases from Press Wood and Laminated Wood Products Manufacturing* (EPA/560/4–88/004i). Washington, DC: EPA.

306. ——. 1993. *Guides to Pollution Prevention: Wood Preserving Industry* (EPA/625/R-93/014). Washington, DC: Office of Research & Development.

307. ———. 1995. *Profile of the Lumber and Wood Products Industry* (EPA/310/R-95/006). Washington, DC: EPA.

READINGS IN HEALTH AND SAFETY

308. Alwis, U., et al. 1999. Dust Exposures in the Wood Processing Industry. *AIHA Journal* 60(5): 641–646.

309. Cote, J., et al. 1991. Occupational Asthma Caused by Exposure to Neurospora in a Plywood Factory Worker. *British Journal of Industrial Medicine* 48(4): 279–282.

310. Drummon, I. 1985. Chlorophenol Exposure in Sawmills. *AIHA Journal* 46(7): B-12.

311. Dutkiewicz, J., et al. 2001a. Exposure to Airborne Microorganisms in Polish Sawmills. *Annals of Agricultural Environmental Medicine* 8: 71–80.

312. ———, et al. 2001b. Response of Sawmill Workers to Work-related Airborne Allergens. *Annals of Agricultural Environmental Medicine* 8: 81–90.

313. Eriksson, K. A., et al. 1996. Terpene Exposure and Respiratory Effects among Sawmill Workers. *Scandinavian Journal of Work & Environmental Health* 22(3): 182–190.

314. Halpin, D. M. G., et al. 1994. Respiratory Symptoms, Immunological Responses, and Aeroallergen Concentrations at a Sawmill. *Occupational and Environmental Medicine* 51: 165–172.

315. Hessel, P. A., et al. 1995. Lung Health in Sawmill Workers Exposed to Pine and Spruce. *Chest* 108: 642–646.

316. Hryhorczuk, D. O., et al. 1998. A Morbidity Study of Former Pentachlorophenol-Production Workers. *Environmental Health Perspectives* 106(7): 401–408.

317. IARC. 1995. *IARC Monographs on the Evaluation of Carcinogenic Risks to Humans: Wood Dust and Formaldehyde* (Monograph No. 62). Lyon, France.

318. Kilburn, K. H. 2000. Indoor Air Effects after Building Renovation and in Manufactured Homes. *American Journal of Medical Science* 320(4): 249–254.

319. Personick, M. E. and Daley, J. R. 1989. Profiles in Safety and Health: Work Hazards of Mobile Homes. *Monthly Labor Review* 112(7): 15–20.

320. Rask-Andersen, A., et al. 1994. Inhalation Fever and Respiratory Symptoms in the Trimming Department of Swedish Sawmills. *American Journal of Industrial Medicine* 25: 65–67.

322: Paper Manufacturing
3221: Pulp, Paper, and Paperboard Mills
3222: Converted Paper Product Manufacturing

READINGS IN HISTORY

321. Berry, S. W. 1998. When Mail Was Armor: Envelopes of the Great Rebellion 1861–1865. *Southern Cultures* 4(3): 63–83.

322. Bowman, B. 2001. *First of a Kind: The Building of the South's First Newsprint Mill.* Lufkin: Best of East Texas.

323. ———. 2002. Southland Paper Mills Inc.: The South's Pioneer Newsprint Manufacturer. *East Texas Historical Journal* 40(2): 16–33.

324. Brand, H. and Huffstutler, C. 1980. The Paper and Plastic Bag Industry: Two Distinct Productivity Phases. *Monthly Labor Review* 103(5): 26–30.

325. Hull, J. P. 1998. Strictly by the Book: Textbooks and the Control of Production in the North American Pulp and Paper Industry. *History of Education* 27(1): 85–95.

326. Ma, H-O. 1992. Substitution Possibilities for Raw Materials in the U.S. Paper and Paperboard Industry. Dissertation, University of Washington.

327. Marino, S. R. 1998. Reading the Envelope: A Literary and Cultural History. *Connecticut Review* 20(2): 17–32.

328. McMurran, B. S. 1989. The Evolution of Market Structure in the North American Newsprint Paper Industry, 1870–1970. Dissertation, University of California, Riverside.

329. Nickel, J. 1993. Stationer's Crests: A Catalog of More Than 200 Embossed Paper Marks 1835–1901. *Manuscripts* 45(3): 199–216.

330. Pikl, J. 1968. Pulp and Paper and Georgia: The Newsprint Paradox. *Forest History* 12(3): 6–19.

331. Reed, G. M. 1995. Realization of a Dream: Charles H. Herty and the South's First Newsprint Mill. *Forest & Conservation History* 39(1): 4–16.

READINGS IN HAZARDOUS WASTE MANAGEMENT AND POLLUTION PREVENTION

332. Beghin, J. and Potier, M. 1997. Effects of Trade Liberalization on the Environment in the Manufacturing Sector. *World Economy* 20(4): 435–456.

333. Blanco, A., et al. 2004. The Challenges of Sustainable Papermaking. *Environmental Science & Technology* 38(21): 414A–420A.

334. Hill, M., et al. 2002. The Greening of a Pulp and Paper Mill: International Paper's Androscoggin Mill, Jay, Maine. *Journal of Industrial Ecology* 6(1): 107–120.

335. Pineda-Henson, R., et al. 2002. Evaluating the Environmental Performance of Pulp and Paper Manufacturing Using the Analytic Hierarchy Process and Life Cycle Assessment. *Journal of Industrial Ecology* 6(1): 15–28.

336. Servos, M. R. 1996. *Environmental Fate and Effects of Pulp and Paper Mill Effluents*. Delray Beach, FL: St. Lucie Press.

337. Sittig, M. 1977. *Pulp and Paper Manufacture: Energy Conservation and Pollution Prevention.* Park Ridge, NJ: Noyes Data.

338. Skinner, H. J. 1939. Waste Problems in the Pulp and Paper Industry. *Industrial & Engineering Chemistry* 31(11): 1331–1335.

339. Sonnenfeld, D. A. 2002. Social Movements and Ecological Modernization: The Transformation of Pulp and Paper Manufacturing. *Development and Change* 33(1): 1–28.

340. Thompson, G., et al. 2001. The Treatment of Pulp and Paper Mill Effluent: A Review. *Bioresource Technology* 77(33): 275–286.

341. U.S. Environmental Protection Agency. 1993. *Handbook on Pollution Prevention Opportunities for Bleached Kraft Pulp and Paper Mills* (EPA/600/R-93/098). Washington, DC: EPA.

READINGS IN HEALTH AND SAFETY

342. Bergmark, G. and Meding, B. 1983. Allergic Contact Dermatitis from Newsprint Paper. *Contact Dermatitis* 9(4): 330.

343. Chan-Yeung, M., et al. 1994. Persistent Asthma after Repeated Exposure to High Concentrations of Gases in Pulp Mills. *American Journal of Respiratory & Critical Care Medicine* 49: 1676–1680.

344. Duchaine, C., et al. 2000. Assessment of Particulates and Bioaerosols in Eastern Canadian Sawmills. *AIHA Journal* 61: 727–732.

345. Glindmeyer, H. W., et al. 2003. Relationship of Asthma to Irritant Gas Exposures in Pulp and Paper Mills. *Respiratory Medicine* 97(5): 541–548.

346. Goyer, N. 1990. Evaluation of Occupational Exposure to Sulfur Compounds in Paper Pulp Kraft Mills. *AIHA Journal* 51(7): 390–394.

347. Papa, G., et al. 1996. Allergic Respiratory Diseases and Environmental Pollution: Experience in the Printing/Paper Manufacturing Industry. *Allergy* 51(11): 833–836.

348. Robinson, C. F., et al. 1986. Mortality among Production Workers in Pulp and Paper Mills. *Scandinavian Journal of Work & Environmental Health* 12(6): 552–560.

349. Solet, D., et al. 1989. Patterns of Mortality in Pulp and Paper Workers. *Journal of Occupational Medicine* 31: 627–630.

350. Teschke, K., et al. 1999. Occupational Exposure to Chemical and Biological Agents in the Nonproduction Departments of Pulp, Paper and Paper Product Mills: An International Study. *AIHA Journal* 60: 73–83.

351. Toren, K., et al. 1996. Health Effects of Working in Pulp and Paper Mills. *American Journal of Industrial Medicine* 29(2): 123–130.

323: Printing and Related Support Activities

READINGS IN HISTORY

352. Armstrong, R. D. 1994. The Matter of Printing: Public Printing in the Western Territories of the United States. *Journal of Government Information* 21(1): 37–47.

353. Auchincloss, K. 2001. The Second Revival: Fine Printing since World War II. *Printing History* 21(1): 3–11.

354. Bellefeuille, J. 1975. Printing in California, 1831–1930. *Pacific Historian* 19(3): 261–270.

355. Benton, M. M. 1997. The Elite Book in America: Fine Book Publishing, 1920–1932. Dissertation, University of California, Berkeley.

356. Conkwright, P. J. 1995. A History of American Printing since 1850. *Princeton University Library Chronicle* 56(2): 259–300.

357. Hennessy, F. 1987. Printing and Pirates: The Genesis of the U.S. Government Printing Office. *Government Publications Review* 14(1): 103–111.

358. Horvitz, E. F. 1990. Harry S. Beck and Other Jewish Printers. *Rhode Island Jewish Historical Notes* 10(4): 494–502.

359. Hudak, L. M. 1978. *Early American Women Printers and Publishers, 1639–1820*. Metuchen, NJ: Scarecrow.

360. Johnston S. P. 1996. *Paper, Presses and Profits: A History of the E.O. Painter Printing Company*. DeLeon Springs, FL: Painter Printing.

361. Kilgour, F. G. 1998. *The Evolution of the Book*. New York: Oxford University Press.

362. Lee, J. B. 2002. Type to Print: The Book and the Type Specimen Book. *Printing History* 22(1–2): 22–71.

363. Pretzer, W. S. 1986. The Printers of Washington, D.C. 1800–1880: Work Culture, Technology and Trade Unionism. Dissertation, Northern Illinois University.

364. Romano, F. 2003. The Phototypesetting Era. *Printing History* 23(2): 36–55.

365. Sanders, M. E. 1980. Wooden Printing Press Building in America: The 1760s to the Middle 19th Century. Dissertation, University of Maryland.

366. Scott, D. T. 1986. Technological Change and Printing Industry Unions, 1958–1983. Dissertation, New School for Social Research.

367. Shields, R. R. 1982. 20th Century American Fine Printing. *Quarterly Journal of the Library of Congress* 39(4): 224–241.

368. Smith D. 1979. Printing: The Trade That Founded a Republic. *American History Illustrated* 13(10): 10–21.

369. Tatham, D. 1971. The Pendleton-Moore Shop: Lithographic Artists in Boston, 1825–1840. *Old Time New England* 62(2): 29–46.

370. ——. 1995. The Lithographic Workshop, 1825–50. *Proceedings of the American Antiquarian Society* 105(1): 71–78.

371. Walters, J. S. 1996. The Pre-superhighway Politics of U.S. Government Printing and Publishing, 1917–1960. *Journal of Government Information* 23(2): 93–121.

READINGS IN HAZARDOUS WASTE MANAGEMENT AND POLLUTION PREVENTION

372. Alabama Department of Environmental Management. 1997. *Offset Printing Waste Reduction Tips*. Montgomery, AL: Pollution Prevention Unit.

373. New York State Department of Environmental Conservation. 1997a. *Environmental Compliance & Pollution Prevention Guide for Lithographic Printers*. Albany, NY: Pollution Prevention Unit.

374. ——. 1997b. *Environmental Self-Assessment for Lithographic Printers*. Pollution Prevention Unit. http://www.dec.ny.gov/docs/permits_ej_operations_pdf/esalith.pdf (accessed 17 October 2008).

375. U.S. Environmental Protection Agency. 1990. *Guides to Pollution Prevention: The Commercial Printing Industry* (EPA/625/7–90/008). Washington, DC: Office of Research & Development.

376. ——. 1994. *Printing Industry and Use Cluster Profile* (EPA/744/R-94/003). Washington, DC: EPA.

377. ——. 1995. *Profile of the Printing and Publishing Industry* (EPA/310/R-95/014). Washington, DC: Office of Enforcement & Compliance Assurance.

378. ——. 1996. *Design for the Environment Flexography Project: Focusing on Flexo Inks* (EPA/744/F-95/006). Washington, DC: EPA.

379. ——. 1997. *Cleaner Technologies Substitutes Assessment: Lithographic Blanket Washes* (EPA/744/R-97/006). Washington, DC: EPA.

380. ——. 1998. *RCRA in Focus: Printing* (EPA/530/K-97/007). Washington, DC: Office of Solid Waste & Emergency Response.

381. ——. 2000. *Flexographic Ink Options* (EPA/744/R-00/004A&B). Washington, DC: EPA.

382. Washington State Department of Ecology. 1996a. *A Guide for Screen Printers* (94–137). http://www.ecy.wa.gov/pubs/94137.pdf (accessed 17 October 2008).

383. ——. 1996b. *A Guide for Lithographic Printers* (94–139). http://www.ecy.wa.gov/pubs/94139.pdf (accessed 17 October 2008).

READINGS IN HEALTH AND SAFETY

384. Cullen, M. R., et al. 1983. Bone Marrow Injury in Lithographers Exposed to Glycol Ethers and Organic Solvents Used in Multicolor Offset and Ultraviolet Curing Printing Processes. *Archives of Environmental Health* 38: 347–354.

385. Horstman, S. W., et al. 2001. Solvent Exposures in Screen Printing Shops. *Journal of Environmental Science & Health A* 36(10): 1957–1973.

386. IARC. 1996. *IARC Monographs on the Evaluation of Carcinogenic Risks to Humans: Printing Processes and Printing Inks, Carbon Black and Some Nitro Compounds* (Monograph No. 65). Lyon, France.

387. White, R. F., et al. 1995. Neurobehavioral Effects of Acute and Chronic Mixed Solvent Exposure in the Screen Printing Industry. *American Journal of Industrial Medicine* 28(2): 221–231.

324: Petroleum and Coal Products Manufacturing

READINGS IN HISTORY

388. Biggs, R. L. 1953. A Geographic Study of the Robinson, Illinois Petroleum Refinery. Dissertation, Southern Illinois University.

389. Bradford, P. A. 1975. *Fragile Structures: A Story of Oil Refineries, National Security, and the Coast of Maine*. New York: Harper's Magazine Press.

390. Brown, E. R., et al. 1994. *Historical Development of Asphalt Content Determination by the Ignition Method* (Report 95–2). Auburn, AL: National Center for Asphalt Technology.

391. Davis, R. L. 1988. *Large Stone Mixes: A Historical Insight*. Riverdale, MD: National Asphalt Pavement Association.

392. Enos, J. L. 1958. A Measure of the Rate of Technological Progress in the Petroleum Refining Industry. *The Journal of Industrial Economics* 6(3): 180–197.

393. Holley, I. B. 2003. Blacktop: How Asphalt Paving Came to the Urban United States. *Technology and Culture* 44(4): 703–733.

394. Maples, R. E. 1993. *Petroleum Refinery Process Economics*. Tulsa, OK: PennWell Books.

395. Nagarajan, K. V. 1990. Petroleum. In *Manufacturing: A Historiographical and Bibliographical Guide,* ed. D. O. Whitten, 173–210. Westport, CT: Greenwood Press.

396. Para, F. R. 2004. *Oil Politics: A Modern History of Petroleum*. New York: I.B. Tauris.

READINGS IN HAZARDOUS WASTE MANAGEMENT AND POLLUTION PREVENTION

397. Berman, E. and Bui, L. T. M. 1998. Environmental Regulation and Productivity: Evidence from Oil Refineries. *NBER Working Paper Series* 676.

398. Edoigiawerie, C. and Spickett, J. 1995. The Environmental Impact of Petroleum in the Environment. *African Journal of Health Sciences* 2(2): 269–276.

399. Epstein, L. N. 1994. *A Review of Pollution Prevention Strategies for Petroleum Refineries*. Washington, DC: National Petroleum Refiners Association.

400. Gammelgard, P. N. 1972. Water Pollution Control in Petroleum Refineries in the United States. *Pure Applied Chemistry* 29(1): 47–56.

401. Hill, J. B. 1939. Waste Problems in the Petroleum Industry. *Industrial & Engineering Chemistry* 31(11): 1361–1363.

402. Kandhal, P. S. 1993. Waste Materials in Hot-Mix Asphalt—An Overview. *ASTM Special Technical Publication No. 1193.*

403. U.S. Department of Energy. 1998. *Energy and Environmental Profile of the U.S. Petroleum Refining Industry*. Washington, DC: Office of Industrial Technologies.

404. Wolf, F. 2005. Resource Availability, Commitment and Environmental Reliability and Safety: A Study of Petroleum Refineries. *Journal of Contingencies and Crisis Management* 13(1): 2–11.

READINGS IN HEALTH AND SAFETY

405. Blot, W. J., et al. 1977. Cancer Mortality in U.S. Counties with Petroleum Industries. *Science* 198 (4312): 51–53.

406. Burstyn, I., et al. 2002. Characterizing Workers Exposure to Bitumen during Hot Mix Paving and Asphalt Mixing Operations. *AIHA Journal* 63(3): 293–299.

407. CONCAWE. 1995. *Health Aspects of Petroleum Fuels* (Report No. 85/51). The Hague, Netherlands: Oil Companies' European Organization for Environmental and Health Protection. Document no. PB85-230480.

408. Cooper, S. P., et al. 1997. Cancer Mortality among Petroleum Refinery Workers in Texas. *Journal of Environmental Pathology & Toxicology* 16(1): 1–14.

409. Cottle, M. K. and Guidotti, T. L. 1990. Process Chemicals in the Oil and Gas Industry: Potential Occupational Hazards. *Toxicology and Industrial Health* 6(1): 41–56.

410. Delzell, E., et al. 1988. Epidemiologic Studies of the Petroleum Industry. *Occupational Medicine* 3(3): 455–474.

411. Edwards, V. C., et al. 1979. Toxicosis Related to the Petroleum Industry. *Veterinary & Human Toxicology* 21(5): 328–337.

412. Gardner, R. 2003. Overview and Characteristics of Some Occupational Exposures and Health Risks on Offshore Oil and Gas Installations. *Annals of Occupational Hygiene* 47(3): 201–210.

413. Harrington, J. M. 1987. Health Experience of Workers in the Petroleum Manufacturing and Distribution Industry: A Review of the Literature. *American Journal of Industrial Medicine* 12(5): 475–497.

414. Heikkila, P., et al. 2002. Occupational Exposure to Bitumen during Road Paving. *AIHA Journal* 63(2): 156–165.

415. IARC. 1989a. *IARC Monographs on the Evaluation of Carcinogenic Risks to Humans: Occupational Exposures in Petroleum Refining; Crude Oil and Major Petroleum Fumes* (Monograph No. 45). Lyon, France.

416. ——. 1989b. *Occupational Exposures in Petroleum Refining: Crude Oil and Major Petroleum Fumes* (Vol. 45). Lyon, France.

417. King, R. W. 1988. Petroleum: Its Composition, Analysis and Processing. *Occupational Medicine* 3(3): 409–430.

418. Kriech, A. J., et al. 2004. Evaluation of Worker Exposure to Asphalt Roofing Fumes. *Journal of Occupational & Environmental Hygiene* (2): 88–98.

419. Luginaah, I. N., et al. 2000. A Longitudinal Study of the Health Impacts of a Petroleum Refinery. *Social Science Medicine* 50(7–8): 1155–1166.

420. Mantel, N. 1985. Exposure of Oil Refinery Workers to the Lubricating-Dewaxing Process. *Journal of the National Cancer Institute* 75(1): 180–181.

421. Nelson, N. A., et al. 1987. Mortality in a Recent Oil Refinery Cohort. *Journal of Occupational Medicine* 29: 610–612.

422. Personick, M. E. 1990. Profiles in Safety and Health: Roofing and Sheet Metal Work. *Monthly Labor Review* 113(9): 27–32.

423. Ruszczak, Z., et al. 1981. Skin Disease in Petrochemical Industry Workers. *Przegl. Derm.* 68: 435–439.

424. Savitz, D. A. and Moure, R. 1984. Cancer Risk among Oil Refinery Workers: A Review of Epidemiologic Studies. *Journal of Occupational Medicine* 26(9): 662–670.

425. Seymour, F. K. and Henry, J. A. 2001. Assessment and Management of Acute Poisoning by Petroleum Products. *Human Experimental Toxicology* 20(11): 551–562.

426. Tsai, S. P., et al. 1991. Morbidity Patterns among Employees at a Petroleum Refinery. *Journal of Occupational Medicine* 33(10): 1076–1080.

427. Vaananen, V., et al. 2005. Dermal Exposure to Polycyclic Aromatic Hydrocarbons among Road Pavers. *Annals of Occupational Hygiene* 49(2): 167–178.

428. Verma, D. K., et al. 2000. Benzene and Total Hydrocarbon Exposures in the Upstream Petroleum Oil and Gas Industry. *AIHA Journal* 61(2): 255–263.

429. ——. 2001. Benzene and Total Hydrocarbon Exposures in the Downstream Petroleum Industries. *AIHA Journal* 62(2): 176–194.

430. Viau, C., et al. 1987. A Cross-sectional Survey of Kidney Function in Refinery Employees. *American Journal of Industrial Medicine* 11: 177–187.

325: Chemical Manufacturing
3251: Basic Chemical Manufacturing
3252: Resin, Synthetic Rubber, and Artificial Synthetic Fibers and Filaments Manufacturing
3253: Pesticide, Fertilizer, and Other Agricultural Chemical Manufacturing
3254: Pharmaceutical and Medicine Manufacturing
3255: Paint, Coating, and Adhesive Manufacturing
3256: Soap, Cleaning Compound, and Toilet Preparation Manufacturing
3259: Other Chemical Product and Preparation Manufacturing

READINGS IN HISTORY

431. Blackford, M. G. 1996. B.F. Goodrich Becomes a Multinational Corporation, 1910–1929. *Essays in Economic & Business History* 14: 365–373.

432. Blakey, A. F. 1973. *The Florida Phosphate Industry: A History of the Development and Use of a Vital Mineral.* Cambridge, MA: Harvard University Press.

433. Bosso, C. J. 1987. *Pesticides and Politics: The Life Cycle of a Public Issue.* Pittsburgh, PA: University of Pittsburgh Press.

434. Bryden, D. J. 1988. Alva Mason, the Franklin Institute and the Origins of Philosophical and Chemical Instrument Manufacture in the Unites States. *Proceedings of the American Philosophical Society* 132(4): 400–419.

435. Castaneda, C. and Pratt J. 1989. New Markets, Outmoded Manufacturing: The Transition from Manufactured Gas to Natural Gas by Northeastern Utilities after World War II. *Business & Economic History* 18: 238–247.

436. Comanor, W. S. 1986. The Political Economy of the Pharmaceutical Industry. *Journal of Economic Literature* 24(3): 1178–1217.

437. Duke, D. R. 1993. Loe Christensen: A Biography of the Founder of Ethyl Alcohol in the United States. Dissertation, Saint Louis University.

438. Fischbaum, M. N. 1990. Drugs. In *Manufacturing: A Historiographical and Bibliographical Guide,* ed. D. O. Whitten, 146–160. Westport, CT: Greenwood Press.

439. Gambardella, A. 1991. Science and Innovation in the U.S. Pharmaceutical Industry during the 1980s. Dissertation, Stanford University.

440. Gunter, V. J. 1995. Environmental Controversies, News Media, and the State: The Case of Synthetic Organic Pesticides in the 1940s, 1950s and 1960s. Dissertation, Michigan State University.

441. Hardin, A. L. 1990. Industry Structure and the Marketing of Synthetic Fibers. *Business & Economic History* 19: 213–222.

442. Harrington J. 1996. The Midwest Agricultural Chemical Association: A Regional Study of an Industry on the Defensive. *Agricultural History* 70(2): 415–438.

443. Harrison, C. S. 2001. Institutional Choice: How the U.S. Research Pharmaceutical Industry Snatched International Victory from the Jaws of Domestic Defeat. Dissertation, University of North Carolina, Chapel Hill.

444. Hensley, J. 1986. Selling the Blast: DuPont's Agricultural Explosives Advertising, 1902–1920. *Delaware History* 22(2): 99–110.

445. Herbert, V. and Attilio, B. 1985. *Synthetic Rubber: The Project That Had to Succeed*. Westport, CT: Greenwood Press.

446. Horton, J. 1977. Phosphate. *Idaho Heritage* 7: 44–47.

447. Jensen, R. E. 1987. Nebraska's World War I Potash Industry. *Nebraska History* 68(1): 28–42.

448. Kadiyali, V. 1994. Pricing and Advertising Strategies for Entry and Accommodation: The Case of the U.S. Photographic Film Industry. Dissertation, Northwestern University.

449. Kirking, D. M. 1983. New Horizons in Pharmaceutical Technology. *Annals of the American Academy of Political & Social Sciences* 468: 182–195.

450. Leo, J. 1996. Change Agents and the Evolution of the U.S. Pharmaceutical Industry, 1920–1960. Dissertation, University of Pennsylvania.

451. Liebenau, J. 1987. *Medical Science and Medical Industry: The Formation of the American Pharmaceutical Industry*. Baltimore, MD: Johns Hopkins University Press.

452. Loughheed, A. L. 1996. Technological Advance in the Manufacture of Chemicals: The Case of Cyanide, 1888–1930. *History of Technology* 18: 81–94.

453. McKinley, S. W. 2003. The Origins of King Phosphate in the New South. Dissertation, University of Delaware.

454. Morris, P. J. 1989. *The American Synthetic Rubber Research Program*. Philadelphia: University of Pennsylvania Press.

455. Oldani, M. J. 2002. Tales from the Script: An Insider/Outside View of Pharmaceutical Sales Practices. *Kroeber Anthropological Society Papers* 87: 147–176.

456. Parascandola, J. 1995. The Emergence of Pharmaceutical Science. *Pharmacy in World History* 37(2): 68–75.

457. Perkins, J. H. 1982. *Insects, Experts and the Insecticide Crisis: The Quest for New Pest Management Strategies*. New York: Plenum.

458. Shepard, C. U. 1893. The Development and Extent of the Fertilizer Industry. *Journal of the American Chemical Society* 15(6): 321–343.

459. Shick, T. W. and Doyle, D. H. 1985. The South Carolina Phosphate Boom and the Stillbirth of the New South, 1867–1920. *South Carolina Historical Magazine* 86(1): 1–31.

460. Simmons, S. A. 1985. American and British Pharmaceutical Industries. *International Social Science Review* 60(3): 72–79.

461. Solo, R. 1959. *Across the High Technology Threshold: The Case of Synthetic Rubber.* Norwood, PA: Norwood Editions.

462. Swan, J. P. 1995. The Evolution of the American Pharmaceutical Industry. *Pharmacy in History* 37(2): 76–86.

463. Vinikas, V. 1992. *Soft Soap, Hard Sell: American Hygiene in an Age of Advertisement*. Ames: Iowa State University Press.

464. Walton, A. L. 1990. Meat and Meat Products. In *Manufacturing: A Historiographical and Bibliographical Guide,* ed. D. O. Whitten, 137–145. Westport, CT: Greenwood Press.

465. Ward, M. R., and Dranove, D. 1995. The Vertical Chain of Research and Development in the Pharmaceutical Industry. *Economic Inquiry* 33(1): 70–87.

466. Wass, A. B. 1992. Natural and Synthetic Dye Use for Protein Fibers in the American Textile Industry, 1856–1891. Dissertation, University of Maryland, College Park.

467. Wells, E. R. 1991. Growing Green: Changing Paradigms in America's Pesticide Use. Dissertation, Bowling Green State University.

468. Wilder, P. S. 1980. The Productivity Trend in the Soaps and Detergents Industry. *Monthly Labor Review* 103(2): 26–30.

469. Wilkinsin, N. B. 1984. *Lammont du Pont and the American Explosives Industry, 1850–1884.* Charlottesville: University of Virginia Press.

470. Wilson, M. R. 2003. Gentlemanly Price-fixing and Its Limits: Collusion and Competition in the U.S. Explosives Industry during the Civil War Era. *Business History Review* 77(2): 207–234.

471. Wines, R. A. 1985. *Fertilizer in America: From Waste Recycling to Resource Exploitation.* Philadelphia: Temple University Press.

472. Woodham, R. T. 2004. History and Development of Engineering Resins. *Polymer Engineering & Science* 25(8): 446–452.

473. Woods, W. G. 1994. An Introduction to Boron: History, Sources, Uses, and Chemistry. *Environmental Health Perspectives* 102(Supp. 7): http://www.ehponline.org/realfiles/members/1994/Suppl-7/woods-full.html (accessed 21 October 2008).

READINGS IN HAZARDOUS WASTE MANAGEMENT AND POLLUTION PREVENTION

474. Abdel-Gawad, S. and Abdel-Shafy, M. 2002. Pollution Control of Industrial Wastewater from Soap and Oil Industries. *Water Science Technology* 46(4–5): 77–82.

475. Bouveng, H. O. 1972. Control of Mercury in Effluents from Chlorine Plants. *Pure Applied Chemistry* 29(1): 75–91.

476. Butner, R. S. 1999. A Heuristic Design Advisor for Incorporating Pollution Prevention Concepts in Chemical Process Design. *Clean Production Processes*. 1(3): 164–169.

477. Hillis, D. R. and Davis, A. D. n.d. *Waste Reduction Strategies for Fiberglass Fabricators.* Greenville, NC: East Carolina University, School of Industry & Technology.

478. McGucken, W. 1991. *Biodegradable: Detergents and the Environment.* College Station: Texas A&M University Press.

479. Platt, B. 1997. *Weaving Textile Reuse into Waste Reduction*. Washington, DC: Institute for Self-Reliance.

480. Post, E. S. 1996. The Importance of State and Plant Characteristics in Determining the Environmental Compliance Costs of Chemical Manufacturing Plants: Evidence from the PACE Survey, 1979–1990. Dissertation, University of Maryland, College Park.

481. Travis, A. S. 2002. Contaminated Earth and Water: A Legacy of the Synthetic Dyestuff Industry. *Ambix* 49(1): 21–50.

482. U.S. Department of Energy. 2000. *Energy and Environmental Profile of the U.S. Chemical Industry*. Washington, DC: Office of Industrial Technologies.

483. U.S. Environmental Protection Agency. 1990. *Guides to Pollution Prevention: The Paint Manufacturing Industry* (EPA/625/7–90/005). Washington, DC: EPA.

484. ——. 1995a. *Profile of the Inorganic Chemical Industry* (EPA/310/R-95/004). Washington, DC: EPA.

485. ——. 1995b. *Profile of the Organic Chemical Industry* (EPA/310/R-95/012). Washington, DC: EPA.

486. ——. 1996a. *Chemical Industry Baseline Compliance Report* (EPA/305/R-96/002). Washington, DC: EPA.

487. ——. 1996b. *Pollution Prevention in the Paints and Coatings Industry* (EPA/625/R-96/003). Washington, DC: Office of Research & Development.

488. ——. 1997a. *Process Based Self-Assessment Tool for the Organic Chemical Industry* (EPA/305/B-97/002). Washington, DC: EPA.

489. ——. 1997b. *Profile of the Pharmaceutical Manufacturing Industry* (EPA/310/R-97/005). Washington, DC: EPA.

490. ——. 1997c. *Profile of the Plastic Resin and Manmade Fibers Industries* (EPA/310/R-97/006). Washington, DC: Office of Enforcement & Compliance Assurance.

491. ——. 1998. *Chemical Industry Compliance Improvement Tool* (EPA/305/B-98/010). Washington, DC: Office of Enforcement & Compliance Assurance.

492. ——. 2000. *Profile of the Agricultural Chemical, Pesticide & Fertilizer Industry* (EPA/310/R-00/003). Washington, DC: Compliance Assistance & Sector Programs Division.

493. ——. 2004. *Compliance Assistance Tool for Pharmaceutical Production, Pesticide Active Ingredient Production, and Miscellaneous Organic Chemical Manufacturing NESHAP: Comparison of Regulatory Requirements and Case Study of Compliance Illustrations for Non-dedicated Equipment* (EPA/305/B-04/001). Washington, DC: Office of Enforcement & Compliance Assurance.

494. Washington Department of Ecology. 2002. *Paint and Coatings Manufacture Sector: A Pollution Prevention Assessment and Guidance*. http://www.ecy.wa.gov/pubs/98410.pdf (accessed 21 October 2008).

READINGS IN HEALTH AND SAFETY

495. Ahlborg, G., et al. 1988. Mutagenic Activity and Metabolites in the Urine of Workers Exposed to Trinitrotoluene. *British Journal of Industrial Medicine* 45(5): 353–358.

496. Amoateng-Adjepong, Y., et al. 1995. Mortality among Workers at a Pesticide Manufacturing Plant. *Journal of Occupational & Environmental Medicine* 37(4): 471–478.

497. Barbone, F., et al. 1992. A Case-controlled Study of Lung Cancer at a Dye and Resin Manufacturing Plant. *American Journal of Industrial Medicine* 22(6): 835–849.

498. Cowels, S. R., et al. 1991. Medical Surveillance for Leukemia at a Petrochemical Manufacturing Complex. *Journal of Occupational Medicine* 33(7): 808–812.

499. DeSilva, P. E. and Donnan, M. B. 1981. Chronic Cadmium Poisoning in a Pigment Manufacturing Plant. *British Journal of Industrial Medicine* 38(1): 76–86.

500. DiFeo, T. J. 2004. Safety and Efficacy: The Role of Chemistry, Manufacturing, and Controls in Pharmaceutical Drug Development. *Drug Development & Industrial Pharmacy* 30(3): 247–257.

501. Foo, S. C., et al. 1994. Chronic Neurobehavioral Effects in Paint Formulators Exposed to Solvents and Noise. *Annals of the Academy of Medicine Singapore* 23: 650–654.

502. Ford, D. P., et al. 1991. A Quantitative Approach to the Characterization of Cumulative and Average Solvent Exposure in Paint Manufacturing Plants. *AIHA Journal* 52: 226–234.

503. Glass, D. C., et al. 1994. Retrospective Assessment of Solvent Exposure in Paint Manufacturing. *Occupational & Environmental Medicine* 51(9): 617–625.

504. Hearne, F. T. and Pifer, J. W. 1999. Mortality Study of Two Overlapping Cohorts of Photographic Film Base Manufacturing Employees Exposed to Methylene Chloride. *Journal of Occupational & Environmental Medicine* 41(12): 1154–1169.

505. Hogstedtc, C. and Andersson, K. 1979. A Cohort Study on Mortality among Dynamite Workers. *Journal of Occupational Medicine* 21(8): 553–556.

506. Hogstedtc, C. and Davidsson, B. 1980. Nitroglycol and Nitroglycerine Exposure in a Dynamite Industry, 1958–1978. *AIHA Journal* 41(5): 373–375.

507. Holmes, T. M., et al. 1986. A Mortality Study of Employees at a Synthetic Rubber Manufacturing Plant. *American Journal of Industrial Medicine* 9(4): 355–362.

508. Hostynek, J. J. and Maibach, H. I. 1988. Chromium in U.S. Household Bleach. *Contact Dermatitis* 18(4): 206–209.

509. IARC. 1977a. *IARC Monographs on the Evaluation of Carcinogenic Risks to Humans: Some Fumigants, the Herbicides 2,4–D and 2,4,5–T, Chlorinated*

Dibenzodioxins and Miscellaneous Industrial Chemicals (Monograph No. 15). Lyon, France.

510. ——. 1977b. *IARC Monographs on the Evaluation of Carcinogenic Risks to Humans: Some Miscellaneous Pharmaceutical Substances* (Monograph No. 11). Lyon, France.

511. ——. 1980. *IARC Monographs on the Evaluation of Carcinogenic Risks to Humans: Some Pharmaceutical Drugs* (Monograph No. 24). Lyon, France.

512. ——. 1982a. *IARC Monographs on the Evaluation of Carcinogenic Risks to Humans: The Rubber Industry* (Monograph No. 28). Lyon, France.

513. ——. 1982b. *IARC Monographs on the Evaluation of Carcinogenic Risks to Humans: Some Industrial Chemicals and Dyestuffs* (Monograph No. 29). Lyon, France.

514. ——. 1990a. *IARC Monographs on the Evaluation of Carcinogenic Risks to Humans: Some Pharmaceutical Drugs* (Monograph No. 50). Lyon, France.

515. ——. 1990b. *IARC Monographs on the Evaluation of Carcinogenic Risks to Humans: Some Pharmaceutical Drugs* (Monograph No. 66). Lyon, France.

516. Kerr, S. M., et al. 2002. Personal Sampling for Inhalable Aerosol Exposures of Carbon Black Manufacturing Industry Workers. *Applied Occupational Environmental Hygiene* 17(10): 681–692.

517. Macaluso, M., et al. 1996. Leukemia and Cumulative Exposure to Butadiene, Styrene and Benzene among Workers in the Synthetic Rubber Industry. *Toxicology* 113(1–3): 190–202.

518. McMichael, A. J., et al. 1975. Solvent Exposure and Leukemia among Rubber Workers. *Journal of Occupational Medicine* 17: 234–239.

519. Olsen, G. W., et al. 1994. Half-century of Cause-specific Mortality Experience of Chemical Manufacturing Employees. *American Journal of Industrial Medicine* 26(2): 203–219.

520. OSHA. 1986. *Safety & Health Guide for the Chemical Industry* (OSHA 3091). Washington, DC: GPO.

521. Ott, M.G., et al. 1985. Mortality among Employees Engaged in Chemical Manufacturing and Related Activities. *American Journal of Epidemiology* 122(2): 311–322.

522. ——. 1989. Lymphatic and Hematopoietic Tissue Cancer in a Chemical Manufacturing Environment. *American Journal of Industrial Medicine* 16(6): 631–643.

523. Pohlova, H. and Sram, R. J. 1985. Cytogenic Analysis of Peripheral Blood Lymphocytes of Workers Occupationally Exposed to Styrene. *Journal of Hygiene Epidemiology & Microbial Immunology* 29(2): 155–161.

524. Sarlo, K. 2003. Control of Occupational Asthma and Allergy in the Detergent Industry. *Annals of Allergy Asthma & Immunology* 90(Supp. 5): 32–34.

525. Sathiakumar, N., et al. 2004. A Cross-sectional Study of Triallate Exposure and Neurological Health among Workers at a Pesticide Manufacturing and Formulating Facility. *Occupational & Environmental Medicine* 61(11): 936–944.

526. Schwartz, B. S., et al. 1990. Solvent-associated Decrements in Olfactory Function in Paint Manufacturing Workers. *American Journal of Industrial Medicine* 18: 697–706.

527. Scott, A .J. 2003. Occupational Health in the Pharmaceutical Industry: An Overview. *Occupational Medicine* 53: 354–356.

528. Seeber, A., et al. 1996. In Search of Dose-Response Relationships of Solvent Mixtures to Neurobehavioral Effects in Paint Manufacturing and Painters. *Food Chemistry & Toxicology* 34(11–12): 1113–1120.

529. Tsai, P. J., et al. 2002. Characterization of PAHs in the Atmosphere of Carbon Black Manufacturing Workplaces. *Journal of Hazardous Materials* 91(1–3): 25–42.

530. Vanhanen, M., et al. 2000. Risk of Enzyme Allergy in the Detergent Industry. *Occupational & Environmental Medicine* 57(2): 121–125.

531. Velagaleti, R., et al. 2002. Impact of Current Good Manufacturing Practices and Emission Regulations and Guidance on the Discharge of Pharmaceutical Chemicals into the Environment from Manufacturing, Use and Disposal. *Environmental Health Perspectives* 110(3): 213–220.

532. Wang, J. D., et al. 1988. Occupational Asthma Due to Toluene Diisocyanate among Velcro-like Tape Manufacturing. *America Journal of Industrial Medicine* 14(1): 73–78.

533. Winchester, R.V. 1985. Solvent Exposure of Workers during Printing Ink Manufacture. *Annals of Occupational Hygiene* 29(4): 517–519.

326: Plastics and Rubber Products Manufacturing

READINGS IN HISTORY

534. Blaszczyk, R. L. 1997. From Invention to Metaphor: Plastics and American Culture. *American Quarterly* 49(3): 640–649.

535. Elias, H-G. 1993. *An Introduction to Plastics*. New York: Wiley.

536. French, M. 1986. Structural Change and Competition in the U.S. Tire Industry, 1920–1937. *Business History Review* 60(1): 28–54.

537. ——. 1987. The Emergence of U.S. Multinational Enterprise: The Goodyear Tire & Rubber Company, 1910–1939. *Economic History Review* 40(1): 64–79.

538. Fricke, E. B. 1982. The New Deal and the Modernization of Small Business: The McCreary Tire & Rubber Company, 1930–1940. *Business History Review* 56(4): 559–576.

539. Friedel, R. 1989. Crazy about Rubber. *American Heritage of Invention & Technology* 5(3): 44–49.

540. Geer, W. C. and Bedford, C. W. 1925. The History of Organic Accelerators in the Rubber Industry. *Industrial & Engineering Chemistry* (April): 393.

541. Gibbons, W. A. 1939. The Rubber Industry, 1839–1939. *Industrial & Engineering Chemistry* (October): 1199.

542. Litz, D. and Moore, L. 1989. Multifactor Productivity Advances in the Tires and Inner Tubes Industry. *Monthly Labor Review* 112(6): 19–27.

543. Meikle, J. L. 1995. *American Plastic: A Cultural History*. New Brunswick, NJ: Rutgers University Press.

544. Moran, T. 2001. The Radial Revolution. *American Heritage of Invention & Technology* 16(4): 28–33.

545. Morris, P. J. T. 1989. *The American Synthetic Rubber Research Program*. Philadelphia: University of Pennsylvania Press.

546. Mossman S. T. 2000. *Early Plastics: Perspectives, 1850–1950*. London: Continuum.

547. Nelson, D. 1987. Mass Production and the U.S. Tire Industry. *Journal of Economic History* 47(2): 329–339.

548. Parkinson, R. 1994. Plating on Plastics: An Industry Review. *Plating & Surface Finishing* 81(99): 32–35.

549. Pennock, P. 1997. The National Recovery Administration and the Rubber Tire Industry, 1933–1935. *Business History Review* 71(4): 543–568.

550. Raeburn, J. 1997. Just One Word: Plastics. *Reviews in American History* 25(1): 140–145.

551. Richardson, T. L. and Lokensgard, E. 2004. *Industrial Plastics: Theory and Applications*. Clifton Park, NY: Thompson Learning.

552. Scala, E. P. 1995. A Brief History of Composites in the U.S.—The Dream and the Success. *JOM* 48(2): 45–48.

553. Seymour, R. B. 1988. History of Touch Plastics with Rubber Inclusion. *Rev. Plast. Mod.* 55(380): 261–265.

554. Seymour, R. B. and Deanin, R. D., eds. 1987. *History of Polymeric Composites*. The Netherlands: VNU Science Press.

555. Solo, R. 1959. *Synthetic Rubber*. Norwood, PA: Norwood Editions.

556. Sull, D. N. 1999. The Dynamics of Standing Still: Firestone Tires and Rubber and the Radial Revolution. *Business History Review* 73(3): 430–464.

557. Walker, A. 1994. Plastics: The Building Blocks of the 20th Century. *Construction History* 10: 67–88.

READINGS IN HAZARDOUS WASTE MANAGEMENT & POLLUTION PREVENTION

558. U.S. Environmental Protection Agency. 1992. *Estimating Chemical Releases from Rubber Production and Compounding* (EPA/560/4–88/004q). Washington, DC: Office of Pesticides & Toxic Substances.

559. ——. 1995. *Profile of the Rubber and Plastics Industry* (EPA/310/R-95/016). Washington, DC: Office of Enforcement & Compliance Assurance.

560. ——. 1997a. *Plastics Materials, Synthetic Resins and Nonvulcanizable Elastomers Manufacturing* (EPA/305/S-97/004). Washington, DC: EPA.

561. ——. 1997b. *Profile of the Plastic Resin & Manmade Fiber Industries* (EPA/310/R-97/006). Washington, DC: EPA.

READINGS IN HEALTH AND SAFETY

562. Chaiear, N. 2001. *Health and Safety in the Rubber Industry*. Shropshire, UK: Rapra Technology.

563. Cummings, B. J. and Booth, K. S. 2002. Industrial Hygiene Sampling for Airborne TDI in Six Flexible Slab Stock Foam Manufacturing Facilities in the U.S. *Applied Occupational & Environmental Hygiene* 17(12): 863–871.

564. DeGesero, R. A. 1974. The Evaluation and Control of Chemicals in Polystyrene Manufacturing. *Annals of Occupational Hygiene* 17(2): 123–129.

565. Delzell, E., Andjelkovich, D. and Tyroler, H. A. 1982. A Case-control Study of Employment Experience and Lung Cancer among Rubber Workers. *American Journal of Industrial Medicine* 3(4):393–404.

566. Dost, A. A., et al. 2000. Exposure to Rubber Fumes and Rubber Process Dust in the General Rubber Goods, Tire Manufacturing and Retread Industries. *Annals of Occupational Hygiene* 44(5): 329–342.

567. Heitbrink, W. A. and McKinnery, W. N., Jr. 1986. Control of Air Contaminants at Mixers and Mills Used in Tire Manufacturing. *AIHA Journal* 47(6):312–321.

568. Holmes, T. M., Buffler, P. A., Holguin, A. H. and Hsi, B. P. 1986. A Mortality Study of Employees at a Synthetic Rubber Manufacturing Plant. *American Journal of Industrial Medicine* 9(4):355–362.

569. IARC. 1979. *IARC Monographs on the Evaluation of Carcinogenic Risks to Humans: Some Monomers, Plastics and Synthetic Elastomers, and Acrolein* (Monograph No. 19). Lyon, France.

570. ——. 1986. *IARC Monographs on the Evaluation of Carcinogenic Risks to Humans: Some Chemicals Used in Plastics and Elastomers* (Monograph No. 39). Lyon, France.

571. Kromhout, H., Swuste, P., Boleij, J. S. 1994. Empirical Modeling of Chemical Exposure in the Rubber-manufacturing Industry. *Annals of Occupational Hygiene* 38(1):3–22.

572. Lewis, R. 1999. Health Issues in Plastics Production and Processing. *Occupational Medicine* 14(4): 777–796.

573. Meikle, J. L. 1997. Material Doubts: The Consequences of Plastic. *Environmental History* 2(3): 278–300.

574. Oury, B., Limasset, J. C. and Protois, J. C. 1997. Assessment of Exposure to Carcinogenic N-nitrosamines in the Rubber Industry. *International Archives of Occupational & Environmental Health*. 70(4): 261–271.

575. Prince, M. M, Ward, E.M, Ruder, A.M., Salvan, A. and Roberts, D.R. 2000. Mortality among Rubber Chemical Manufacturing Workers. *American Journal of Industrial Medicine* 37(6): 590–598.

576. Slovak, A. J. 1984. Occupational Hazards of Polyethylene and Polypropylene Manufacture. *Progress in Clinical Biological Research* 141: 309–312.

577. Straif, K., Chambless, L., Weiland, S. K., Wienke, A., Bungers, M., Taeger, D. and Keil, U. 1999. Occupational Risk Factors for Mortality from Stomach and Lung Cancer among Rubber Workers: An Analysis Using Internal Controls and Refined Exposure Assessment. *International Journal of Epidemiology* 28(6): 1037–1043.

578. Toeppen-Sprigg, B. 1999. Management of Dermatitis in the Rubber Manufacturing Industry. *Occupational Medicine* 4(4): 797–818.

579. Vainio, H., et al. 1980. Chemical Hazards in the Plastics Industry. *Journal of Toxicology & Environmental Health* 6(5–6): 1179–1186.

580. Vermeulen, R., de Hartog, J., Swuste, P. and Kromhout, H. 2001. Trends in Exposure to Inhalable Particulate and Dermal Contamination in the Rubber Manufacturing Industry: Effectiveness of Control Measures Implemented over a Nine-year Period. *Annals of Occupational Hygiene* 44(5): 343–354.

581. Vermeulen, R., Heideman, J., Bos, R. P. and Kromhout, H. 2000. Identification of Dermal Exposure Pathways in the Rubber Manufacturing Industry. *Annals of Occupational Hygiene* 44(7): 533–541.

582. Vermeulen, R., Kromhout, H., Bruynzeel, D. P., de Boer, E.M. and Brunekreef, B. 2001. Dermal Exposure, Hand Washing, and Hand Dermatitis in the Rubber Manufacturing Industry. *Epidemiology* 12(3): 350–354.

583. Veys, C. A. 2004. Bladder Tumors in Rubber Workers: A Factory Study 1946–1995. *Occup Medicine* 54(5): 322–329.

584. Williams, T. M., Harris, R. L., Arp, E. W., Symons, M. J. and Van Ert, M.D. 1980. Worker Exposure to Chemical Agents in the Manufacture of Rubber Tires and Tubes: Particulates. *AIHA Journal* 41(3):204–211.

585. Yip, E. and Cacioli, P. 2002. The Manufacture of Gloves from Natural Rubber Latex. *Journal of Allergy & Clinical Immunology* 110(Suppl. 2): S3–S14.

327: Nonmetallic Mineral Product Manufacturing
3271: Clay Product and Refractory Manufacturing
3272: Glass and Glass Product Manufacturing
3273: Cement and Concrete Product Manufacturing
3274: Lime and Gypsum Product Manufacturing
3279: Other Nonmetallic Mineral Product Manufacturing

READINGS IN HISTORY

586. Bair, T. 1967. Industry Relocation and Restrictive Work Practices: The Flat Glass Industry. *Land Economics* 43(1): 96–100.

587. Bogue, R. H. 1955. *The Chemistry of Portland Cement*. New York: Reinhold Publishing.

588. Crispin, S. and Hollerbach, M. 1987. Grand Ledge Clay Product Co. *Michigan History* 71(4): 24–29.

589. Diehl, M. W. and Diehl, A. C. 2001. Economics, Ideology and the Brick Industry in Tucson. *Journal of the Southwest* 43(3): 423–446.

590. Gallaway, L. E. 1962. The Origin and Early Years of the Federation of Flat Glass Workers of America. *Labor History* 3(1): 92–102.

591. Geary, K. 1983. Gypsum Production at Blue Diamond, Nevada, 1924–1959. *Nevada Historical Society Quarterly* 26(2): 111–121.

592. Grindle, R. L. 1971. *Quarry and Kiln: The Story of Maine's Lime Industry*. Rockland, ME: Courier-Gazette.

593. Hood, G. 1972. *Bonnin and Morris of Philadelphia: The First American Porcelain Factory, 1770–1772*. Chapel Hill: University of North Carolina Press.

594. Keane, L. A. 1916. Plaster of Paris. *Journal of the American Chemical Society* 20(8): 701–723.

595. Meade, R. K. 1906. A Review of the American Portland Cement Industry. *Journal of the American Chemical Society* 28(9): 1257–1264.

596. Measell, J. S. 1980. The Indiana Tumbler & Goblet Company, 1894–1903. *Indiana Magazine of History* 76(4): 319–333.

597. Newlong, H. 1976. *A Selection of Historic American Papers on Concrete, 1876–1926*. Detroit: American Concrete Institute.

598. Palmer, A. 1976. Glass Production in 18th Century America: The Wistarburgh Enterprise. *Winterthur Portfolio* 11: 75–101.

599. Pilling, R. W. 1979. Henry Chapman Mercer and the Moravian Tile Works. *American Art & Antiques* 2(6): 78–83.

600. Rauschenberg, B. L. 1991. Brick and Tile Manufacturing in the South Carolina Low Country, 1750–1800. *Journal of Early Southern Decorative Arts* 17(2): 103–113.

601. Renk, N. F. 1992. Back to Basics: The Lake Pend Oreille Lime and Cement Industry and Its Regional Impact. *Idaho Yesterdays* 36(1): 2–11.

602. Rolando, V. R. 1991. *200 Years of Soot and Sweat: The History and Archeology of Vermont's Iron, Charcoal, and Lime Industries*. Manchester: Vermont Archeological Society.

603. Spillman, J. S. 1983. Pressed-glass Designs in the United States and Europe. *Magazine Antiques* 124(1): 130–139.

604. Stern, M. J. 2003. Industrial Structure and Occupational Health: The American Pottery Industry, 1897–1929. *Business History Review* 77(3): 417.

605. Walters, W. D. 1979. Abandoned 19th Century Brick and Tile Works in Central Illinois: An Introduction from Local Sources. *Industrial Archeology Review* 4(1): 70–80.

READINGS IN HAZARDOUS WASTE MANAGEMENT AND POLLUTION PREVENTION

606. Alexander, L. D. 1995. Waste Disposal Options for Refractory Tear Out. *American Ceramic Society Bulletin* 74(1): 68.

607. Andereola, N. M., et al. 1993. Recycling of Ceramic Wastes in Tile Bodies to Reduce Pollution. *American Ceramic Society Bulletin* 72(4): 65.

608. Campbell, I. 1998. Lead and Cadmium Free Glasses and Frits. *Glass Technology* 39(2): 38–41.

609. Chandra, S. 1997. *Waste Materials Used in Concrete Manufacturing*. Norwich, NY: William Andrew.

610. D'Anna, G. 2003. Best Available Techniques for the Heavy Clay Industry. *Industrial Ceramics* 991: 16–24.

611. Environment Agency. 2001a. *Integrated Pollution Prevention and Control: Guidance for the Cement and Lime Sector* (IPPC S3.01). Bristol, UK: The Environment Agency.

612. ———. 2001b. *Integrated Pollution Prevention and Control: Guidance for the Glass Manufacturing Sector* (A1 Processes). IPPC S3.03. Bristol, UK: The Environment Agency.

613. Fuessle, R. W. and Taylor, M. A. 2000. Stabilization of Arsenic and Barium Rich Glass Manufacturing Waste. *Journal of Environmental Engineering* 126(3): 272–278.

614. Gorokhovsky, A. V., et al. 2004. Composite Materials Based on Wastes of Flat Glass Processing. *Waste Management* 25(7): 733–736.

615. Navarro, J. E. E. 1998. Integrated Pollution Prevention and Control: The Example of the Ceramic Tile Industry. *Industrial Ceramics* 939: 452–464.

616. Tarbeev, V. V. 1998. Diminishing Atmospheric Pollution in Production of Polished Glass. *Glass and Ceramics* 55(1/2): 33–35.

617. Uchikawa, H. and Hanehara, S. 1997. *Waste Materials and Industrial By-Product in Concrete Manufacturing*. Park Ridge, NJ: Noyes Publications.

READINGS IN HEALTH AND SAFETY

618. Andersson, L., et al. 1990. Some Hygienic Observations from the Glass Industry. *International Archives of Occupational & Environmental Health*. 62: 249–252.

619. Apostoli, P., et al. 1998. Multiple Exposure to Arsenic, Antimony and Other Elements in Art Glass Manufacturing. *American Journal of Industrial Medicine* 34(1): 65–72.

620. Bazroy, J., et al. 2003. Magnitude and Risk Factors of Injuries in a Glass Bottle Manufacturing Plant. *Journal of Occupational Health* 45(1): 53–59.

621. Braun, S. R. and Tsiatis, A. 1979. Pulmonary Abnormalities in Art Glassblowers. *Journal of Occupational Medicine* 21(7): 487–489.

622. Fishmann, B. B. and Kuznetsov, V. A. 2001. Cytotoxicity of High-clay Refractory Dust. *Med. Tr. Prom. Ekol.* 2: 35–38.

623. Goh, C. L. and Gan, S. L. 1996. Change in Cement Manufacturing Process: A Cause for Decline in Chromate Allergy? *Contact Dermatitis* 34(1): 51–54.

624. IARC. 1994. *IARC Monographs on the Evaluation of Carcinogenic Risks to Humans: Beryllium, Cadmium, Mercury, and Exposures in the Glass Manufacturing Industry* (Monograph No. 58). Lyon, France.

625. International Finance Corporation. 1998. *Environmental, Health and Safety Guidelines for Ceramic Tile Manufacturing.*

626. Klein, M. 1970. Lead Release from Earthenware. *NEJM* 283(23): 1292.

627. LeMasters, G. K., et al. 2003. Mortality of Workers Occupationally Exposed to Refractory Ceramic Fibers. *Journal of Occupational & Environmental Medicine* 45(4): 440–450.

628. Ludersdorf, R., et al. 1987. Biological Assessment of Exposure to Antimony and Lead in the Glass Producing Industry. *International Archives of Occupational & Environmental Health* 59: 469–474.

629. Mansberger, F. 2001. 19th Century Pottery Production in Illinois. *Material Culture* 33(1): 46–67.

630. Molina-Ballesteros, G., et al. 1980. Lead Exposure in Two Pottery Handicraft Populations. *Archives of Investigative Medicine* 11(1): 147–154.

631. Morley, S. E., et al. 1996. Cement Related Burns. *Burns* 22(8): 646–647.

632. Sankila, R., et al. 1990. Cancer Risk among Glass Factory Workers: An Excess of Lung Cancer? *British Journal of Industrial Medicine* 47: 815–818.

633. Stern, M. J. 2003. Industrial Structure and Occupational Health: The American Pottery Industry, 1897–1929. *Business History Review* 77(3): 417–445.

634. Wilson, R., et al. 1999. A Risk Assessment for Exposure to Glass Wool. *Regulatory Toxicology & Pharmacology* 30(Part 1): 96–109.

635. Wingren, G. and Axelson, O. 1987. Mortality in the Swedish Glassworks Industry. *Scandinavian Journal of Work & Environmental Health* 13: 412–416.

636. ——. 1992. Cluster of Brain Cancers Spuriously Suggesting Occupational Risk among Glassworkers. *Scandinavian Journal of Work & Environmental Health* 18: 85–89.

637. Zuskin, E., et al. 1998. Respiratory Findings in Workers Employed in the Brick-Manufacturing Industry. *Journal of Occupational Medicine* 40(9): 814–820.

331: Primary Metal Manufacturing
3311: Iron and Steel Mills and Ferroalloy Manufacturing
3312: Steel Product Manufacturing from Purchased Steel
3313: Alumina and Aluminum Production and Processing
3314: Nonferrous Metal (Except Aluminum) Production and Processing
3315: Foundries

READINGS IN HISTORY

638. Abbott, C. M. 1965. Early Copper Smelting in Vermont. *Vermont History* 33(1): 233–242.

639. Carey, J. L. 1973. Productivity in the Steel Foundries Industry. *Monthly Labor Review* 96(5): 8–11.

640. Eichenwald, K. 1985. America's Successful Steel Industry. *Washington Monthly* 17(1): 40–44.

641. Harris, H. J. 2000. The Rocky Road to Mass Production: Change and Continuity in the U.S. Foundry Industry, 1890–1940. *Enterprise & Society* 1(2): 391–437.

642. Ingham, J. N. 1991. *Making Iron and Steel: Independent Mills in Pittsburgh, 1820–1920.* Columbus: Ohio State University Press.

643. McFadden, J. M. 1978. Monopoly in Barbed Wire: The Formation of the American Steel and Wire Company. *Business History Review* 52(4): 465–489.

644. Patton, S. G. 1990. Blast Furnaces and Steel Mills. In *Manufacturing: A Historiographical and Bibliographical Guide,* ed. D. O. Whitten, 239–262. Westport, CT: Greenwood Press.

645. Quivik, F. L. 1999. Smoke and Tailings: An Environmental History of Copper Smelting Technologies in Montana: 1880–1930. Dissertation, University of Pennsylvania.

646. Schatzberg, E. 2003. Symbolic Culture and Technological Change: The Cultural History of Aluminum as an Industrial Material. *Enterprise & Society* 4(2): 226–271.

647. Young, O. E. 1983. Origins of the American Copper Industry. *Journal of the Early Republic* 3(2): 117–137.

READINGS IN HAZARDOUS WASTE MANAGEMENT AND POLLUTION PREVENTION

648. Environment Agency. 2002a. *Integrated Pollution Prevention and Control: Guidance for the Coke, Iron and Steel Sector* (IPPC S2.01). Bristol, UK: The Environment Agency.

649. ——. 2002b. *Integrated Pollution Prevention and Control: Technical Guidance for the Non-Ferrous Metals and the Production of Carbon and Graphite* (IPPC S2.02). Bristol, UK: The Environment Agency.

650. International Aluminum Institute. 2003. *Life Cycle Assessment of Aluminum: Inventory Data for the Worldwide Primary Aluminum Industry*. London: International Aluminum Institute.

651. Kotzin, E. L. 1992. *Steel Foundries: Air Pollution Engineering Manual*. New York: Van Nostrand Reinhold.

652. Misurovic, A. 2003. Management of Drinking, Surface and Underground Water Polluted by Waste from Aluminum Industry. *Critical Reviews in Analytical Chemistry* 33: 311–319.

653. North American Die Casting Association. 1996. *Pollution Prevention Practices for the Die Casting Industry*. Rosemont, IL: North American Die Casting Association.

654. Swain, R. E. 1939. Waste Problems in the Nonferrous Smelting Industry. *Industrial & Engineering Chemistry* 31(11): 1358–1361.

655. Teworte, W. 1972. Measures Taken against Water Pollution in Basic Nonferrous Metal Industries. *Pure Applied Chemistry* 29(1): 235–261.

656. U.S. Department of Energy. 1997. *Energy and Environmental Profile of the U.S. Aluminum Industry.* Washington, DC: Office of Industrial Technologies.

657. ——. 1999. *Energy and Environmental Profile of the U.S. Metal Casting Industry.* Washington, DC: Office of Industrial Technologies.

658. ——. 2000. *Energy and Environmental Profile of the U.S. Steel Industry.* Washington, DC: Office of Industrial Technologies.

659. U.S. Environmental Protection Agency. 1981. *Summary of Factors Affecting Compliance by Ferrous Foundries* (EPA/340/1–80/020). Washington, DC: Office of General Enforcement.

660. ———. 1992. *Guides to Pollution Prevention: The Metal Casting & Heat Treating Industry* (EPA/625/R-92/009). Washington, DC: Office of Research & Development.

661. ———. 1995. *Profile of the Iron and Steel Industry* (EPA/310/R-95/005). Washington, DC: EPA.

662. ———. 1997. *Profile of the Metal Casting Industry* (EPA/310/R-97/004). Washington, DC: Enforcement & Compliance Assurance.

663. ———. 1998. *Pollution Prevention in the Primary Metals Industry* (EPA/749/B-98/003). Washington, DC: EPA.

READINGS IN HEALTH AND SAFETY

664. Abramson, M. J., et al. 1989. Does Aluminum Smelting Cause Lung Disease? *American Review of Respiratory Disease* 139(4): 1042–1057.

665. Benke, G., et al. 1998. Exposures in the Alumina and Primary Aluminum Industry: An Historical Review. *Annals of Occupational Hygiene* 42(3): 173–189.

666. Burton, D. J. 1980. An Overview of Occupational Exposure Control Technology in Nonferrous Smelting. *American Journal of Industrial Medicine* 1(3–4): 293–305.

667. Chaurel, C., et al. 1993. Environmental Stresses and Strains in an Extreme Situation: The Repair of Electrometallurgy Furnaces. *International Archives of Occupational & Environmental Health* 65(4): 253–258.

668. Enterline, P. E. and Marsh, G. M. 1980. Mortality Studies of Smelter Workers. *American Journal of Industrial Medicine* 1(3–4): 251–259.

669. Hansen, K. S. 1983. Occupational Accidents in Iron Foundries. *Ugeskr. Laeger*. 145(31): 2395–2398.

670. Hong, C. J., et al. 1999. Determination of Urinary Arsenic, Mercury, and Selenium in Steel Production Workers. *Biological & Trace Element Research* 70: 29–40.

671. Kilburn, K. H. 1998. Neurobehavioral Impairment and Symptoms Associated with Aluminum Smelting. *Archives of Environmental Health* 53(5): 329–335.

672. Koskela, R. S. 1994. Cardiovascular Diseases among Foundry Workers Exposed to Carbon Monoxide. *Scandinavian Journal of Work & Environmental Health* 20(4): 286–293.

673. Kuo, H. W., Chang, C. L., Liang, W. M. and Chung, B.C. 1999. Respiratory Abnormalities among Male Foundry Workers in Central Taiwan. *Occupational Medicine* (Lond.) 49(8):499–505.

674. Longstreth, W. T., et al. 1985. Neurologic Disorder in Three Aluminum Smelter Workers. *Archives of Internal Medicine* 145: 1972–1975.

675. Monarca, S., et al. 1984. Mutagenic/carcinogenic Hazards in a Cold-rolling Steel Plant Exposed to Mineral Oils. *International Archives of Occupational & Environmental Medicine* 54(4): 345–354.

676. Palda, V. A. 2003. Is Foundry Work a Risk for Cardiovascular Disease? A Systematic Review. *Occupational Medicine* 53: 179–190.

677. Palmer, W. G. and Scott, W. D. 1981. Lung Cancer in Ferrous Foundry Workers: A Review. *AIHA Journal* 42(5):329–340.

678. Rodriguez, V., Tardon, A., Kogevinas, M., Prieto, C. S., Cueto A., Garcia, M., Menendez, I. A. and Zaplana, J. 2000. Lung Cancer Risk in Iron and Steel Foundry Workers: A Nested Case Control Study in Asturias, Spain. *American Journal of Industrial Medicine* 38(6):644–650.

679. Ronneberg, A. and Langmark, F. 1992. Epidemiologic Evidence of Cancer in Aluminum Reduction Plant Workers. *American Journal of Industrial Medicine* 22(4): 573–590.

680. Schnor, T. M., et al. 1995. Mortality in a Cohort of Antimony Smelter Workers. *American Journal of Industrial Medicine* 27: 759–770.

681. Seixas, N. S., et al. 2000. Urinary Fluoride as an Exposure Index in Aluminum Smelting. *AIHA Journal* 61(1): 89–94.

682. Sherson, D., Svane, O. and Lynge, E. 1991. Cancer Incidence among Foundry Workers in Denmark. *Archives of Environmental Health* 46(2):75–81.

683. Silverstein, M., et al. 1986. Mortality among Ferrous Foundry Workers. *American Journal of Industrial Medicine* 10(1): 27–43.

684. Tola, S. 1980. Epidemiology of Lung Cancer in Foundries. *Journal of Toxicology & Environmental Health* 6(5–6): 1195–1200.

685. Tossavainen, A. 1976. Metal Fumes in Foundries. *Scandinavian Journal of Work & Environmental Health* 2 (Supp. 1): 42–49.

686. ——. 1990. Estimated Risk of Lung Cancer Attributable to Occupational Exposures in Iron and Steel Foundries. *IARC Science Publications* 104: 363–367.

687. Verma, D. K. and Shaw, D. S. 1991. An Evaluation of Airborne Nickel, Zinc and Lead Exposure at Hot Dip Galvanizing Plant. *AIHA Journal* 52(12): 511–515.

688. Westberg, H. B. and Bellander, T. 2003. Epidemiological Adaptation of Quartz Exposure Modeling in Swedish Aluminum Foundries: Nested Case-control Study on Lung Cancer. *Applied Occupational Environmental Hygiene* 18(12):1006–1013.

689. Westberg, H. B. and Selden, A. 1999. Chemical Work in Secondary Aluminum Smelting. *Archives of Environmental Health* 54(5): 364.

690. Westberg, H. B., Selden, A. I. and Bellander, T. 2001. Exposure to Chemical Agents in Swedish Aluminum Foundries and Aluminum Smelting Plants—A Comprehensive Survey. *Applied Occupational Environmental Hygiene* 16(1):66–77.

332: Fabricated Metal Product Manufacturing
3321: Forging and Stamping
3322: Cutlery and Hand Tool Manufacturing
3323: Architectural and Structural Metals Manufacturing
3324: Boiler, Tank, and Shipping Container Manufacturing
3325: Hardware Manufacturing
3326: Spring and Wire Product Manufacturing
3327: Machine Shops; Turned Product; and Screw, Nut, and Bolt Manufacturing
3328: Coating, Engraving, Heat Treating, and Allied Activities
3329: Other Fabricated Metal Product Manufacturing

READINGS IN HISTORY

691. Becker, W. H. 1971. American Wholesale Hardware Trade Associations, 1870–1900. *Business History Review* 45(2): 179–200.

692. Bezis-Selfa, J. 2003. *Forging America: Ironworkers, Adventures and the Industrious Revolution.* Ithaca, N.Y. Cornell University Press.

693. Black, R. M. 1983. *The History of Electric Wires and Cables*. London: Peter Peregrinus Ltd.

694. Brand, H. and Huffstutler, C. 1986. Trends of Labor Productivity in Metal Stamping Industries. *Monthly Labor Review* 109(5): 13–20.

695. Green, A. 1999. Tin Men on Parade: The Art of Sheet-metal Workers. *Labor's Heritage* 10(3): 34–47.

696. Otto, P. F. 1980. Productivity Growth Below Average in Fabricated Structural Metals. *Monthly Labor Review* 103(6): 27–31.

697. Persigehl, E. S. and Olsen, J. G. 1986. Productivity in the Metal Doors, Sash and Trim Industry. *Monthly Labor Review* 109(3): 27–30.

698. Personick, M. E. 1990. Profiles in Safety and Health: Roofing and Sheet Metal Work. *Monthly Labor Review* 113(9): 27–32.

699. Tyler, J. D. 1978. 18th and 19th Century Cast-Iron Cooking Utensils. *Early American Life* 9(2): 29–31.

700. Waters, D. D. 1977. From Pure Coin: The Manufacture of American Silver Flatware, 1800–1860. *Winterthur Portfolio* 12: 19–33.

701. Wheeler, S. E. 2000. The Racine Hardware Manufacturing Company. *Nautical Research Journal* 45(2): 69–82.

READINGS IN HAZARDOUS WASTE MANAGEMENT AND POLLUTION PREVENTION

702. Kansas State University. n.d. *Metal Finishers & Electroplaters: Regulatory Requirements & Pollution Prevention Ideas.*

703. ——. *Pollution Prevention for the Electroplating & Metal Finishing Industry. Small Business Environmental Assistance Program.* http://www.engext.ksu.edu/ppi/publications/manual/MetalFinish/contents.html (accessed 23 October 2008).

704. NSF International. 1998. *Environmental Management Systems: A Guide for Metal Finishers.* Ann Arbor, MI: NSF International.

705. U.S. Environmental Protection Agency. 1989. *Waste Minimization in Metals Parts Cleaning* (EPA/530/SW-89/049). Washington, DC: Office of Solid Waste & Emergency Response.

706. ——. 1993. *Waste Minimization for the Metal Finishing Industry* (EPA/742/B-93/005). Washington, DC: EPA.

707. ——. 1994. *Guide to Cleaner Technologies: Alternative Metal Finishes* (EPA/625/R-4/007). Washington, DC: Office of Research & Development.

708. ——. 1995. *Profile of the Fabricated Metal Products Industry* (EPA/310/R-95/007). Washington, DC: Office of Enforcement & Compliance Assurance.

709. Washington State Department of Ecology. 1999. *Metal Machining Sector Pollution Prevention Assessment & Guidance* (99–412). http://www.ecy.wa.gov/pubs/99412.pdf (accessed 23 October 2008).

READINGS IN HEALTH AND SAFETY

710. Bright, P., et al. 1997. Occupational Asthma Due to Chrome and Nickel Electroplating. *Thorax* 52(1): 28–32.

711. Goldsmith, A. H., et al. 1976. Health Hazards from Oil, Soot and Metals at a Hot Forging Operation. *AIHA Journal* 37(4): 217–226.

712. Kipling, M. D. 1976. Health Hazards and Powder Metallurgy. *Journal of Social & Occupational Medicine* 26(3): 81–84.

713. Makinen, M. and Linnainmaa, M. 2004. Dermal Exposure to Chromium in Electroplating. *Annals of Occupational Hygiene* 48(3): 209–217.

714. Park, R. M. 2001. Medical Insurance Claims and Surveillance for Occupational Disease: Analysis of Respiratory, Cardiac, and Cancer Outcomes in Auto Industry Tool Grinding Operations. *Journal of Occupational and Environmental Medicine* 43(4): 335–346.

715. Sadhra, S., et al. 2002. Worker's Understanding of Chemical Risks: Electroplating Case Study. *Occupational & Environmental Medicine* 59(10): 689–695.

716. Silverstein, M., et al. 1988. Mortality among Bearing Plant Workers Exposed to Metalworking Fluids and Abrasives. *Journal of Occupational Medicine* 30(9): 706–714.

717. Sorahan, T., et al. 1998. Lung Cancer Mortality in Nickel/Chromium Platers. *Occupational and Environmental Medicine* 55: 236–242.

333: Machinery Manufacturing
3331: Agriculture, Construction, and Mining Machinery Manufacturing
3332: Industrial Machinery Manufacturing
3333: Commercial and Service Industry Machinery Manufacturing
3334: HVAC and Commercial Refrigeration Equipment Manufacturing
3335: Metalworking Machinery Manufacturing
3336: Engine, Turbine, and Power Transmission Equipment Manufacturing
3339: Other General Purpose Machinery Manufacturing

READINGS IN HISTORY

718. Ackermann, M. E. 2002. *Cool Comfort: America's Romance with Air Conditioning*. Washington, DC: Smithsonian Institution Press.

719. Berlin, L. R. 2001. Robert Noyce and Fairchild Semiconductor, 1957–1968. *Business History Review* 75(1): 63–101.

720. Cooper, G. 1988. Manufactured Weather: A History of Air Conditioning in the United States, 1902–1955. Dissertation, University of California, Santa Barbara.

721. ——. 1994. Custom Design, Engineering Guarantees and Unpatentable Data: The Air Conditioning Industry, 1902–1935. *Technology and Culture* 35(3): 506–536.

722. ——. 1998. *Air Conditioning America: Engineers and the Controlled Environment, 1900–1960*. Baltimore, MD: Johns Hopkins University Press.

723. Davis, A. B. and Dreyfuss, M. S. 1987. *The Finest Instruments Ever Made: A Bibliography of Medical, Dental, Optical, and Pharmaceutical Company Trade Literature, 1700–1939.* Arlington, MS: Medical History Publishing Association.

724. DiFilippo, A. J. 1981. Militarism and Machine: Tool Building in America. Dissertation, Temple University.

725. Dix, K. 1980. Joseph Joy and His Mobile Loading Machine. *West Virginia History* 41(3): 226–244.

726. Dobis, V. J. 1995. Structure and Rivalry in the U.S. Farm Machinery Industry, 1945–1993. Dissertation, University of Manitoba.

727. Duke, J. 1980. Construction Machinery Industry Posts Slow Rise in Productivity. *Monthly Labor Review* 103(7): 33–36.

728. Englund, J. H. 1978. An Outline of the Development of Wood Moulding Machinery. *APT Bulletin* 10(4): 20–45.

729. Falk, E. and Litz, D. 1991. Multifactor Productivity in Farm and Garden Equipment. *Monthly Labor Review* 114(6): 27–38.

730. Feather, C. E. 1995. Go See Willard: Selling Farm Machinery in Preston County. *Goldenseal* 21(1): 32–37.

731. Goodall, L. E. 1992. The Rotary Power Mower and Its Inventor: Leonard B. Goodall. *Missouri Historical Review* 86(3): 248–264.

732. Gooding, J. 1991. Machine Tool Mecca. *American Heritage of Invention & Technology* 6(3): 42–48.

733. Goodwin, J. 2001. *Otis: Giving Rise to the Modern City*. Chicago: Dee.

734. Graham, R. C. 1990. Farm and Garden Machinery. In *Manufacturing: A Historiographical and Bibliographical Guide,* ed. D. O. Whitten, 265–278. Westport, CT: Greenwood Press.

735. Hannon, J. U. 1982. City Size and Ethnic Discrimination: Michigan Agricultural Implements and Iron Working Industries, 1890. *Journal of Economic History* 42(4): 825–845.

736. Hapgood, F. 2004. The Underground Cutting Edge: The Innovators Who Made Digging Tunnels High-tech. *American Heritage of Invention & Technology* 20(2): 42–48.

737. Haycraft, W. R. 2000. *Yellow Steel: The Story of the Earthmoving Equipment Industry.* Urbana: University of Illinois Press.

738. Herman, A. S. and Ferris, J. W. 1982. Productivity Growth Average in Farm Machinery Manufacturing. *Monthly Labor Review* 105(10): 6–10.

739. Huffstutler, C. and Brand, H. 1984. Productivity in Making Air Conditioners, Refrigeration Equipment and Furnaces. *Monthly Labor Review* 107(12): 11–17.

740. Jackson, D. D. 1989. Elevating Thoughts from Elisha Otis and Fellow Uplifters. *Smithsonian* 20(8): 210–234.

741. Kipnis, S. and Huffstutler, C. 1990. Productivity Trends in the Photographic Equipment and Supplies Industry. *Monthly Labor Review* 113(6): 39–49.

742. Leichtle, K. E. 1995. Power in the Heartland: Tractor Manufacturers in the Midwest. *Agricultural History* 69(2): 314–325.

743. Mazzoleni, R. 1998. The Agency System in the International Distribution of U.S. Machine Tools, 1900–1915. *Business & Economic History* 27(2): 420–430.

744. Meyer, D. R. 1998. Formation of Advance Technology Districts: New England Textile Machinery and Firearms: 1790–1820. *Economic Geography* (special issue): 31–45.

745. OhUallachain, B. 1997. Restructuring the American Semiconductor Industry. *Annals of the Association of American Geographers* 87(2): 217–237.

746. O'Neil, B. A. 1987. The Mining Machinery Industry: Labor Productivity Trends, 1972–1984. *Monthly Labor Review* 110(6): 31–36.

747. Pudup, M. B. 1987. From Farm to Factory: Structuring and Location of the U.S. Farm Machinery Industry. *Economic Geography* 63(3): 203–222.

748. Thomson, R. 1989. Invention, Markets, and the Scope of the Firm: The 19th Century U.S. Shoe Machinery Industry. *Business & Economic History* 18: 140–149.

749. Turner, G. L'E. 1969. The History of Optical Instruments. *History of Science* 8: 53–93.

750. Walker, J. C. 1973. Elisha Otis: The Man Who Took the Walking Out of Going Up. *Mankind* 4(1): 44–49.

751. Welsh, P. C. 1966. The Metallic Woodworking Plane: An American Contribution to Hand Tool Design. *Technology and Culture* 7(1): 38–47.

752. Wheeler, S. E. 2003. Andrew Jackson Pierce: Marine Engine Builder. *Nautical Research Journal* 48(3): 130–138.

753. Winder, G. M. 1995. Before the Corporation and Mass Production: The Licensing Regime in the Manufacture of North American Harvesting Machinery, 1830–1910. *Annals of the Association of American Geographers* 85(3): 521–552.

754. Witzel, M. K. 1997. Gas Pumps. *American Heritage of Invention & Technology* 12(3): 58–63.

READINGS IN HAZARDOUS WASTE MANAGEMENT AND POLLUTION PREVENTION

755. Frazier, G. and John Deere & Co. n.d. *Global Trends in Agriculture: A Farm Machinery Manufacturer's Perspective*. Available at http://www.ngplains.org/documents/Global_Trends_in_Agriculture_Gary_Frazier_.pdf.

756. Harney, J. 2006. Waste Management—Greener Is Smarter. *The Manufacturer* (November).

757. Klassen, R. D. 2000. Exploring the Linkage between Investment in Manufacturing and Environmental Technologies. *International Journal of Operations & Production Management* 20(2): 127–147.

758. Kutz, M., ed. 2007. *Handbook of Farm, Dairy and Food Machinery*. Norwich, NY: William Andrew Publishing.

READINGS IN HEALTH & SAFETY

759. Somerson, S. J. and Sicilia, M. R. 1992. Historical Perspectives on the Development and Use of Mechanical Ventilation. *AANA Journal* 60(1): 83–94.

760. Xiao, G. B., et al. 2004. Study on Musculoskeletal Disorders in a Machinery Manufacturing Plant. *Journal of Occupational & Environmental Medicine* 46(4): 341–346.

334: Computer and Electronic Product Manufacturing
3341: Computer and Peripheral Equipment Manufacturing
3342: Communications Equipment Manufacturing
3343: Audio and Video Equipment Manufacturing
3344: Semiconductor and Other Electronic Component Manufacturing

3345: Navigational, Measuring, Electro Medical, and Control Instruments Manufacturing

3346: Manufacturing and Reproducing Magnetic and Optical Media

READINGS IN HISTORY

761. Angel, D. P. and Engstrom, J. 1995. Manufacturing Systems and Technological Change: The U.S. Personal Computer Industry. *Economic Geography* 71(1): 79–102.

762. Cusumano, M. A. 1991. Factor Concepts and Practices in Software Development. *Annals of the History of Computing* 13(1): 3–32.

763. Einstein, M. E. and Franklin, J. C. 1986. Computer Manufacturing Enters a New Era of Growth. *Monthly Labor Review* 109(9): 9–16.

764. Friberg, E. E. 1990. Electronic Computing Equipment. In *Manufacturing: A Historiographical and Bibliographical Guide,* ed. D. O. Whitten, 279–300. Westport, CT: Greenwood Press.

765. McGahagan, T. A. 1990. Scientific and Measuring Instruments. In *Manufacturing: A Historiographical and Bibliographical Guide,* ed. D. O. Whitten, 358–366. Westport, CT: Greenwood Press.

766. Park, B-G. 1992. A Dynamic Perspective of Strategic Groups and Performance: A Longitudinal Study of the U.S. Computer Equipment Industry. Dissertation, Virginia Polytechnic Institute.

READINGS IN HAZARDOUS WASTE MANAGEMENT AND POLLUTION PREVENTION

767. ACGIH. 1993. *Hazard Assessment and Control Technology in Semiconductor Manufacturing.* Chelsea, MI: Lewis Publishers.

768. Sturgill, J. A., et al. 1999. Pollution Prevention in the Semiconductor Industry through Recovery and Recycling of Gallium and Arsenic from Gas Solid Wastes. *Clean Production Processes* 1(4): 248–256.

769. U.S. Environmental Protection Agency. 1994. *Federal Environmental Regulations Potentially Affecting the Computer Industry* (EPA/744/B-93/002). Washington, DC: EPA.

770. ——. 1995. *Profile of the Electronics & Computer Industry* (EPA/310/R-95/002). Washington, DC: EPA.

771. ——. 1998. *Computer Display Industry & Technology Profile* (EPA/744/R-98/005). Washington, DC: EPA.

772. Van Zant, P. 1990. *Microchip Fabrication: A Practical Guide to Semiconductor Processing*. 2nd ed. New York: McGraw-Hill.

773. Williamson, M. C. 1998. *Energy Efficiency in Semiconductor Manufacturing: A Tool for Cost Savings and Pollution Prevention*. 8th ed. London: Semiconductor Fabtech.

774. Yang, J. and Sieben, C. 2001. Environmental Challenges in Computer Manufacturing. *Environment International* 27(1): 49–59.

READINGS IN HEALTH AND SAFETY

775. Bertazzi, P. A., et al. 1987. Cancer Mortality of Capacitor Manufacturing Workers. *American Journal of Industrial Medicine* 11(2): 165–176.

776. Broadwell, D. K., et al. 1995. Work Site Clinical and Neurobehavioral Assessment of Solvent-exposed Microelectronics Workers. *American Journal of Industrial Medicine* 27: 677–698.

777. Fowler, B. A., et al. 1993. Cancer Risks for Humans from Exposure to the Semiconductor Metals. *Scandinavian Journal of Work & Environmental Health* 19: 101–103.

778. Liao, Y-H., et al. 2004. Biological Monitoring of Exposures to Aluminum, Gallium, Indium, Arsenic and Antimony in Optoelectronic Industry Workers. *JOEM* 46: 931–936.

779. Mallin, K., et al. 2004. Cohort Mortality Study of Capacitor Manufacturing Workers, 1944–2000. *Journal of Occupational & Environmental Medicine* 46(6): 565–576.

780. OSHA. 1988. *Safety & Health Guide for the Microelectronics Industry* (OSHA 3107). http://www.dtic.mil/cgi-bin/GetTRDoc?AD=ADA400601&Location=U2&doc=GetTRDoc.pdf (accessed 23 October 2008).

781. Rischitelli, G. 2005. Dermatitis in a Printed-Circuit Board Manufacturing Facility. *Contact Dermatitis* 52(2): 78–81.

782. Schenker, M. B., et al. 1995. Association of Spontaneous Abortion and Other Reproductive Effects with Work in the Semiconductor Industry. *American Journal of Industrial Medicine* 28: 639–659.

335: Electrical Equipment, Appliance, and Component Manufacturing
3351: Electric Lighting Equipment Manufacturing

3352: Household Appliance Manufacturing
3353: Electrical Equipment Manufacturing
3359: Other Electrical Equipment and Component Manufacturing

READINGS IN HISTORY

783. Carter, S. B. 1990. Household Appliances. In *Manufacturing: A Historiographical and Bibliographical Guide,* ed. D. O. Whitten, 303–308. Westport, CT: Greenwood Press.

784. Hausman, W. J. 1990. Electric Lighting and Wiring Equipment. In *Manufacturing: A Historiographical and Bibliographical Guide,* ed. D. O. Whitten, 309–319. Westport, CT: Greenwood Press.

READINGS IN HAZARDOUS WASTE MANAGEMENT AND POLLUTION PREVENTION

785. U.S. Environmental Protection Agency. 1995. *Printed Wiring Board Industry and Use Cluster Profile* (EPA/744/R-05/005). Washington, DC: EPA.

786. ——. 1998. *Alternative Technologies for Making Holes Conductive: Cleaner Technologies for Printed Wiring Board Manufacturers* (EPA/744/R-98/002). Washington, DC: EPA.

787. ——. 2001a. *Alternative Technologies for Surface Finishing: Cleaner Technologies for Printed Wiring Board Manufacturers* (EPA/744/R-01/001). Washington, DC: EPA.

788. ——. 2001b. *Printed Wiring Board Surface Finishes: Cleaner Technologies Substitutes Assessment, Vol. 1 and Vol. 2* (EPA/744/R-01/003A & B). Washington, DC: EPA.

READINGS IN HEALTH AND SAFETY

789. Dewell, P. 1994. Concentrations of Cadmium in Air and Urine in an Alkaline Battery Works. *Annals of Occupational Hygiene* 38(2): 209–218.

790. Hanley, K. W. and Lenhart, S. W. 2000. Manganese Dioxide Exposures and Respirator Performance at an Alkaline Battery Plant. *Applied Occupational Environmental Hygiene* 15(7): 542–549.

791. Harsh, G. G. 1954. A Study of the Dust, Mold, and Bacteria Content of the Exhaust of Various Types of Vacuum Cleaners: A Preliminary Report. *Annals of Allergy* 12(6): 705–709.

792. Heo, Y., et al. 2004. Serum IgE Elevation Correlates with Blood Lead Levels in Battery Manufacturing Workers. *Human Experimental Toxicology* 23(5): 209–213.

793. LaVecchia, C. and Bosetti, C. 2003. Cancer Risk in Carbon Electrode Workers. *European Journal of Cancer Prevention* 12(5): 431–434.

794. Lemond, J. M. 2005. Case Study of a Specialty Battery Manufacturing Facility. *Journal of Occupational & Environmental Hygiene* 2(3): D8–D12.

795. Tepper, A., et al. 2001. Blood Pressure, Left Ventricular Mass, and Lead Exposure in Battery Manufacturing Workers. *American Journal of Industrial Medicine* 40(1): 63–72.

796. Trakumas, S., et al. 2001. Particle Emission Characteristics of Filter-equipped Vacuum Cleaners. *AIHA Journal* 62(4): 482–493.

797. Vinentini, L., et al. 2002. Comparison of Vacuum Cleaners. *Allergy* 57(6): 555.

336: Transportation Equipment Manufacturing
3361: Motor Vehicle Manufacturing
3362: Motor Vehicle Body and Trailer Manufacturing
3363: Motor Vehicle Parts Manufacturing
3364: Aerospace Product and Parts Manufacturing
3365: Railroad Rolling Stock Manufacturing
3366: Ship and Boat Building
3369: Other Transportation Equipment Manufacturing

READINGS IN HISTORY

798. Aduddell, R. M. and Cain, L. P. 1990. Motor Vehicles and Equipment. In *Manufacturing: A Historiographical and Bibliographical Guide,* ed. D. O. Whitten, 331–344. Westport, CT: Greenwood Press.

799. Bernstein, B. J. 1966. The Automobile Industry and the Coming of the Second World War. *Southwestern Social Science Quarterly* 47(1): 22–33.

800. Boas, C. W. 1961. Location Patterns of American Automobile Assembly Plants. *Economic Geography* 37(3): 218–230.

801. Breihan, J. R. 1997. From Amusements to Weapons: The Glenn L. Martin Aircraft Company of California, 1910–1917. *Journal of the West* 36(3): 29–38.

802. Brooks, S. K. 1975. The Motorcycle in American Culture: From Conception to 1935. Dissertation, George Washington University.

803. Brown, J. K. 1993. The Baldwin Locomotive Works. *Business & Economic History* 22(1): 7–12.

804. Fleming, T. 1994. Tanks. *American Heritage of Invention & Technology* 10(3): 54–63.

805. Froesch, C. 2002. The Fokker Aircraft Corporation of America. *American Aviation Historical Society Journal* 47(3): 216–225.

806. Golich, V. L. 1992. From Competition to Collaboration: The Challenge of Commercial Class Aircraft Manufacturing. *International Organization* 46(4): 899–934.

807. Herligy, D. V. 1992. The Bicycle Story. *American Heritage of Invention & Technology* 7(4): 48–59.

808. Hochfelder, D. and Helper, S. 1996. Suppliers and Product Development in Early American Automobile Industry. *Business & Economic History* 25(2): 39–51.

809. Hounshell, D. A. 2000. Automation, Transfer Machinery, and Mass Production in the U.S. Automobile Industry in the Post–World War II Era. *Enterprise & Society* 1(1): 100–138.

810. Institute for Great Lakes Research. 1988. *American Ship Building Company and Predecessors, 1867–1920*. Perrysburg, OH: Bowling Green State University Press.

811. Johnson, A. 2000. Engineering Culture and the Production of Knowledge: An Intellectual History of Anti Lock Braking Systems. Dissertation, Princeton University.

812. Johnson, S. B. 2004. Launch Vehicles and the Development of Systems Engineering. *Quest* 11(1): 40–50.

813. Kronemer, A. and Henneberger, J. E. 1993. Productivity in Aircraft Manufacturing. *Monthly Labor Review* 116(6): 24–33.

814. Lane, F. C. 1951. *Ships for Victory: A History of Shipbuilding Under the U.S. Maritime Commission in World War II*. Baltimore, MD: Johns Hopkins University Press.

815. Lanigan, C. 2003. The Early Automobile Industry in Southwestern Pennsylvania. *Western Pennsylvania History* 86(4): 30–39.

816. Launius, R. G. and Jenkins D. R, eds. 2002. *To Reach the High Frontier: A History of U.S. Launch Vehicles*. Lexington: University Press of Kentucky.

817. Raff, D. M. G. 1991. Making Cars and Making Money in the Interwar Automobile Industry. *Business History Review* 65(4): 721–753.

818. Reynolds, G. 2004. Built Along the Shores of Macatawa: The History of Boat Building in Holland, Michigan. *Inland Seas* 60(20): 98–108.

819. Rubenstein, J. M. 1988. Changing Distribution of American Motor-Vehicle Parts Suppliers. *Geographical Review* 78(3): 288–298.

820. Schafer, L. S. 1986. A Brief Brilliance: Pennsylvania's Early Automakers. *Pennsylvania Heritage* 12(4): 26–31.

821. Simonson, G. R. 1960. The Demand for Aircraft and the Aircraft Industry, 1907–1958. *Journal of Economic History* 20(3): 361–382.

822. Spitzer, P. 1998. Harsh Ways: Edward N. Heath and the Shipbuilding Trade. *Pacific Northwest Quarterly* 90(1): 3–16.

823. ——. 2004. Boeing as a Startup Company, 1915–1917. *Pacific Northwest Quarterly* 95(3): 140–148.

824. Stanfield, P. 1992. Heritage Design: The Harley-Davidson Motor Company. *Journal of Design History* 5(2): 141–155.

825. Sturdevant, R. W. and Spires, D. N. 1997. Mile High Adventures: Highlights from Colorado Aerospace Industry, 1923–1997. *Journal of the West* 36(3): 67–77.

826. Sullivan, T. E. 1990. Ship and Boat Building and Repairing. In *Manufacturing: A Historiographical and Bibliographical Guide,* ed. D. O. Whitten, 343–354. Wesport, CT: Greenwood Press.

827. Sung, S. 1991. A Quantitative Analysis of the Dual Use of Technology in the U.S. Airframe Manufacturing and Aircraft Engine Industries, 1961–1985. Dissertation, Rensselaer Polytechnic Institute.

828. Thompson, S. L. 2000. The Arts of the Motorcycle: Biology, Culture, and Aesthetics in Technological Choice. *Technology and Culture* 41(1): 99–115.

829. Usselman, S. W. 1984. Air Brakes for Freight Trains: Technological Innovation in the American Railroad Industry, 1869–1900. *Business History Review* 58(1): 30–50.

830. Wheeler, S. E. 2002. The Racine Boat Manufacturing Company of Muskegon. *Nautical Research Journal* 47(2): 67–77.

831. Wik, R. M. 1980. The American Farm Tractor as Father of the Military Tank. *Agricultural History* 54(1): 126–133.

READINGS IN HAZARDOUS WASTE MANAGEMENT AND POLLUTION PREVENTION

832. Horne, D. F. 1986. *Aircraft Production Technology*. Cambridge: Cambridge University Press.

833. Oregon Department of Environmental Quality. 2000. *Best Management Practices for Oregon Shipyards*. Western Region, Water Quality Division. http://www.deq.state.or.us/WQ/pubs/bmps/shipyards.pdf (accessed 23 October 2008).

834. U.S. Environmental Protection Agency. 1997. *Profile of the Shipbuilding and Repair Industry* (EPA/310/R-97/008). Washington, DC: Enforcement & Compliance Assurance.

835. ———. 1998. *Profile of the Aerospace Industry* (EPA/310/R-98/001). Washington, DC: EPA.

836. U.S. Navy. 1994. *Hazardous Waste Minimization Guide for Shipyards*. Washington, DC: EPA.

READINGS IN HEALTH AND SAFETY

837. Boice, J. D., et al. 1999. Mortality among Aircraft Manufacturing Workers. *Occupational and Environmental Medicine* 56(9): 581–597.

838. Brigham, C. R. and Landrigan, P. J. 1985. Safety and Health in Boatbuilding and Repair. *American Journal of Industrial Medicine* 8(3): 169–182.

839. Brown, D. A. and Delzell, E. 2000. Motor Vehicle Manufacturing and Prostate Cancer. *American Journal of Industrial Medicine* 38(1): 59–70.

840. Castillon, K. 2000. Overexposure to Methylene Bisphenyl Isocyanate in a Motor Vehicle Parts Manufacturing Facility. *Applied Occupational & Environmental Hygiene* 15(3): 251–252.

841. Delzell, E. and Monson, R. R. 1984. Mortality among Rubber Workers. VII. Aerospace Workers. *American Journal of Industrial Medicine* 6(4): 265–271.

842. Delzell, E., et al. 1994. Cancer Mortality among Women Employed in Motor Vehicle Manufacturing. *Journal of Occupational Medicine* 36(11): 1251–1259.

843. ———. 2003. Mortality among Hourly Motor Vehicle Manufacturing Workers. *Journal of Occupational Medicine* 45(8): 813–830.

844. Ducatman, A. M., et al. 1986. Germ Cell Tumors of the Testicle among Aircraft Repairman. *Journal of Urology* 136: 834–836.

845. Garabrant, D., et al. 1988. Mortality of Aircraft Manufacturing Workers in Southern California. *American Journal of Industrial Medicine* 13: 683–693.

846. Hussain, T. 2004. Musculoskeletal Symptoms among Truck Assembly Workers. *Occupational Medicine* 54: 506–512.

847. Jagels, R. 1985. Health Hazards of Natural and Introduced Chemical Components of Boatbuilding Woods. *American Journal of Industrial Medicine* 8(3): 241–245.

848. Lilies, R., et al. 1982. Assessment of Lead Health Hazards in a Body Shop of an Automobile Assembly Plant. *American Journal of Industrial Medicine* 3(1): 33–51.

849. Marano, D. E., et al. 2000. Exposure Assessment for a Large Epidemiological Study of Aircraft Manufacturing Workers. *Applied Occupational & Environmental Hygiene* 15(8): 644–656.

850. Morgan, R. W., et al. 1998. Mortality of Aerospace Workers Exposed to Trichloroethylene. *Epidemiology* 9: 424–431.

851. Park, R. M., et al. 1990. Brain Cancer Mortality at a Manufacturer of Aerospace Electromechanical Systems. *American Journal of Industrial Medicine* 17(5): 537–552.

852. Paustenbach, D. J., et al. 2004. Environmental and Occupational Health Hazards Associated with the Presence of Asbestos in Brake Linings and Pads (1900 to Present): A State of the Art Review. *Journal of Toxicology & Environmental Health* B: *Critical Reviews* 7(1): 25–80.

853. Pylypiw, H. M. 1991. Polychlorinated Biphenyls in Auto Parts and Auto Fluff. *Bulletin of Environmental Contamination & Toxicology* 46(5): 681–685.

854. Spirtas, R., et al. 1991. Retrospective Cohort Mortality Study of Workers at an Aircraft Manufacturing Maintenance Facility: 1. Rpidemiological Results. *British Journal of Industrial Medicine* 48: 515–530.

855. Stewart, P. A., et al. 1991. Retrospective Cohort Mortality Study of Workers at an Aircraft Manufacturing Maintenance Facility: 11. Exposures and Their Assessment. *British Journal of Industrial Medicine* 48: 531–537.

856. Swanson, G. M., et al. 1985. Colon Cancer Incidence among Model Makers and Patternmakers in the Automobile Manufacturing Industry. A Continuing Dilemma. *Journal of Occupational Medicine* 27(8): 567–569.

857. Warner, M., et al. 1998. Acute Traumatic Injuries in Automotive Manufacturing. *American Journal of Industrial Medicine* 34(4): 351–358.

858. Wohl, A. R., et al. 1995. Occupational Injury in Female Aerospace Workers. *Epidemiology* 6(2): 110–114.

337: Furniture and Related Product Manufacturing
3371: Household and Institutional Furniture and Kitchen Cabinet Manufacturing
3372: Office Furniture (Including Fixtures) Manufacturing
3379: Other Furniture Related Product Manufacturing

READINGS IN HISTORY

859. Brand, H. and Bennett, N. 1985. Productivity Trends in Kitchen Cabinet Manufacturing. *Monthly Labor Review* 108(3): 24–30.

860. Congram, M. 1980. Haircloth Upholstery. *19th Century* 6(4): 48–50.

861. Cooper, W. 1980. Form and Fabric: The Art of the Upholsterer. *Art & Antiques* 3(2): 110–117.

862. Darling, M. W. 1997. Ambient Modernism: The Domestic Furniture Designs of the George Nelson Office, 1944–1963. Dissertation, University of California, Santa Barbara.

863. Garrett, E. D. 1985. The American Home, Part V: Venetian Shutters and Blinds. *Magazine Antiques* 128(2): 259–265.

864. Grier, K. C. 1989. Culture and Comfort: People, Parlors, and Upholstery, 1850–1930. *American Quarterly* 41(September): 568–576.

865. Henneberger, J. E. 1978. Productivity Growth below Average in the Household Furniture Industry. *Monthly Labor Review* 101(11): 23–29.

866. ———. 1982. The Office Furniture Industry: Patterns in Productivity. *Monthly Labor Review* 105(12): 33–37.

867. Madigan, M. J. S. 1975. The Influence of Charles Locke Eastlake on American Furniture Manufacture, 1870–1890. *Winterthur Portfolio* 10: 1–22.

868. Muth, T. M. and Falk, E. T. 1994. Multifactor Productivity in Household Furniture. *Monthly Labor Review* 117(6): 35–46.

869. Ostergard, D. E., ed. 1987. *Bent Wood and Metal Furniture, 1850–1946*. Seattle: University of Washington Press.

870. Roth, R. 2004. Oriental Carpet Furniture: A Furnishing Fashion in the West in the Late 19th Century. *Studies in the Decorative Arts* 11(2): 25–58.

871. Swaim, C. 2004. The Hoosier Kitchen Cabinet: The Homemaker's Friend. *Indiana & Midwestern History* 16(2): 26–35.

872. Uber, T. L. 2001. Creating the Steel Chapel: A Study of Commercial Office Furniture Design in the United States from 1876–1925. Dissertation, Case Western Reserve University.

873. Veigle, J. and Brand, H. 1982. Millwork Industry Shows Slow Growth in Productivity. *Monthly Labor Review* 105(9): 21–26.

874. Waggoner, E. 1996. A Splendid Job Done: Mattress Making in the Great Depression. *Goldenseal* 22(3): 46–50.

READINGS IN HAZARDOUS WASTE MANAGEMENT AND POLLUTION PREVENTION

875. Bartlett, C. 2003. *Environmental Reporting for Furniture Manufacturers: How to Get Started*. Hertfordshire, UK: FIRA International.

876. Bovea, M. D. and Vidal, R. 2003. Materials Selection for Sustainable Product Design: A Case Study of Wood Based Furniture Eco-design. *Materials & Design* 25(2):111–116.

877. Kohl, J., et al. 1986. *Managing and Recycling Solvents in the Furniture Industry*. Raleigh: North Carolina State University, Nuclear Engineering Department.

878. Tennessee Valley Authority. n.d. *Waste Reduction Guide: Wood Furniture Industries*. Knoxville, TN: TVA.

879. U.S. Environmental Protection Agency. 1995. *Profile of the Wood Furniture and Fixtures Industry* (EPA/310/R-95/003). Washington, DC: EPA.

880. ——. 2004. *RCRA in Focus: Furniture Manufacturing and Refinishing* (EPA/530/K-03/005). Washington, DC: Office of Solid Waste & Emergency Response.

READINGS IN HEALTH AND SAFETY

881. Acheson, E. D. 1976. Nasal Cancer in the Furniture and Boot and Shoe Manufacturing Industries. *Preventive Medicine* 5(2): 295–315.

882. deBoer, R. 2002. Allergens, Der p 1, Der f 1, Fel d 1 and Can f 1, in Newly Bought Mattresses for Infants. *Clinical & Experimental Allergy* 32(11): 1602–1605.

883. Dyson, W. L. 1981. Furniture Manufacturing: A Step-by-Step Profile. *Occupational Health & Safety* 50(7): 23–25.

884. Fleming, P. J., et al. 2002. Mattresses, Microenvironments, and Multivariate Analysis. *British Medical Journal* 325(7371): 981–982.

885. Gan, S. L., et al. 1987. Occupational Dermatitis among Sanders in the Furniture Industry. *Contact Dermatitis* 17(4): 237–240.

886. Hall, R. M. 2003. Exposures to Lead, Metals, and Wood Dust during Stripping and Refinishing Furniture. *Applied Occupational Environmental Hygiene* 18(9): 639–645.

887. Imbus, H. R. and Dyson, W. L. 1987. A Review of Nasal Cancer in Furniture Manufacturing and Woodworking in North Carolina. *Journal of Occupational Medicine* 29(9): 734–740.

888. Krysinska-Traczyk, E., et al. 2002. Exposure to Airborne Microorganisms in Furniture Factories. *Annals of Agricultural & Environmental Medicine* 9(1): 85–90.

889. Miller, B. A., et al. 1994. Extended Mortality Follow-up among Men and Women in a U.S. Furniture Workers Union. *American Journal of Industrial Medicine* 25(4): 537–549.

890. Mirka, G.A. 2005. Development of an Ergonomics Guidelines for the Furniture Manufacturing Industry. *Applied Ergonomics* 36(2): 241–247.

891. Mirka, G. A., et al. 2001. Ergonomic Interventions for the Furniture Manufacturing Industry. *International Journal of Industrial Ergonomics* 29(5): 263–273.

892. Mohtashamipur, E., et al. 1989. Cancer Epidemiology of Woodworking. *Journal of Cancer Research & Clinical Oncology* 115(6): 503–515.

893. Sass-Kortsak, A. M., et al. 1986. Wood Dust and Formaldehyde Exposures in the Cabinet Making Industry. *AIHA Journal* 47(12): 747–753.

894. Schei, M. A., et al. 2002. House-dust Mites and Mattresses. *Allergy* 57(6): 538–542.

895. Siebens, D. P. and Tippens, L. 1995. Occupational Hazards in the Furniture Industry. *North Carolina Medical Journal* 56(5): 202–204.

896. Skorska, C., et al. 2002. Response of Furniture Factory Workers to Work-related Airborne Allergens. *Annals of Agricultural & Environmental Medicine* 9(1): 91–97.

897. Topping, M. D., et al. 1981. Castor Bean Allergy in the Upholstery Department of a Furniture Factory. *British Journal of Industrial Medicine* 38(3): 293–296.

339: Miscellaneous Manufacturing
3391: Medical Equipment and Supplies Manufacturing
3399: Other Miscellaneous Manufacturing

READINGS IN HISTORY

898. Anonymous. 1990. Good Manufacturing Practices, Including Quality Assurance for Dental Materials. *International Dental Journal* 40(4): 253–256.

899. Augustyn, F. J. 2002. The American Switzerland: New England as a Toymaking Center. *Journal of Popular Culture* 36(1): 1–13.

900. Bowdidge, J. S. 1990. Toys and Sporting Goods. In *Manufacturing: A Historiographical and Bibliographical Guide,* ed. D. O. Whitten, 389–406. Westport, CT: Greenwood Press.

901. Bro, P. A. 1993. The Development of the American Made Saxophone. Dissertation, Northwestern University.

902. Cassedy, J. H. 1976. The Microscope in American Medical Science, 1840–1860. *Isis* 67(236): 76–97.

903. Copeley, W. 1991. Musical Instrument Makers of New Hampshire, 1800–1960. *Historical New Hampshire* 46(4): 231–248.

904. Cross, G. 1997. *Kid's Stuff: Toys and the Changing World of American Childhood.* Cambridge, MA: Harvard University Press.

905. Daniels, M. 1980. The Ingenious Pen. *American Archivist* 43(3): 312–324.

906. Davis, A. B. 1978. Historical Studies of Medical Instruments. *History of Science* 16(2): 107–133.

907. Edmondson, J. M. 1997. *American Surgical Instruments: An Illustrated History of Their Manufacture and a Directory of Instrument Makers to 1900.* San Francisco: Norman.

908. Fletcher, N. H. 1999. The Nonlinear Physics of Musical Instruments. *Rep. Prog. Phys.* 62: 723–764.

909. Glinsky, A V. 1992. The Theremin in the Emergence of Electronic Music. Dissertation, New York University.

910. Golder, R. I. 1993. Chicago Surgical Instrument Makers, 1855–1899. *Caducues* 9(2): 99–110.

911. Greer, A. L. 1981. Medical Technology: Assessment, Adoption and Utilization. *Journal of Medical Systems* 5(1–2): 129–145.

912. Groce, N. J. 1982. Musical Instrument Making in New York City during the 18th and 19th Centuries. Dissertation, University of Michigan.

913. Gustatitis, J. 2000. Neon: A Light History. *American History* 35(2): 28–32.

914. Hart, K. W. 1973. Cincinnati Organ Builders of the 19th Century. *Cincinnati Historical Society Bulletin* 31(2): 79–98.

915. Hedley-White, J. and Milamed, D. R. 1999. Equipment Standards: History, Litigation, and Advice. *Annals of Surgery* 230(1): 120.

916. Jakel, J. A. and Sculle, K. A. 2004. *Signs in America's Auto Age*. Iowa City: University of Iowa Pres.

917. Kass, P. and Olmert, M. 1977. Violin Making as American Art. *Smithsonian* 8(6): 106–110.

918. Kay, L. E. 1988. Laboratory Technology and Biological Knowledge: The Tiselius Electrophoresis Apparatus, 1930–1945. *History & Philosophy of the Life Sciences* 10(1): 51–72.

919. Petroski, H. 1990. *The Pencil: A History of Design and Circumstance*. New York: Knopf.

920. Phillips, W. G. and Rila, C. 1987. Individual First Aid Equipment in the U.S. Service. *Military Collector & Historian* 39(2): 50–63.

921. Pinch, T. and Trocco, F. 2002. *Analog Days: The Invention and Impact of the Moog Synthesizer*. Cambridge, MA: Harvard University Press.

922. Rauschenberg, B. L. 1990. Coffin Making and Undertaking in Charleston and Its Environs, 1705–1820. *Journal of Early Southern Decorative Arts* 16(1): 18–63.

923. Repin, V. A. and Valeev, B. A. 1988. Outlook for Growth in the Manufacture of Instruments Used in Oral Medicine. *Med. Tekh.* (Russian). 1: 36–40.

924. Robert, E. B. 1987. *Technological Innovation and Medical Devices*. Cambridge, MA: Sloan School of Management, MIT.

925. Shapiro-Perl, N. R. 1984. Labor Process and Class Relations in the Costume Jewelry Industry. Dissertation, University of Connecticut.

926. Sokal, M. M., et al. 1975. A National Inventory of Historic Psychological Apparatus. *Journal of the History of the Behavioral Sciences* 11(3): 284–286.

927. Venerable, C. L. 1993. Silver in America, 1840–1940: Production, Marketing and Consumption. Dissertation, Boston University.

928. Walsh, M. 1992. Plush Endeavors: An Analysis of the Modern American Soft Toy Industry. *Business History Review* 66(4): 637–670.

929. Worden, G. 1993. Steel Knives and Iron Lungs: Medical Instruments as Medical History. *Caduceus* 9(2): 111–118.

930. Zhang, Y-De., et al. 2002. Robotic System Approach for Complete Denture Manufacturing. *Transactions on Mechatronics* 7(3): 392–396.

READINGS IN HAZARDOUS WASTE MANAGEMENT AND POLLUTION PREVENTION

931. California Department of Toxic Substances Control. 2002. *Jewelry Manufacturing Pollution Prevention Recommendations.* http://www.dtsc.ca.gov/HazardousWaste/Jewelry/upload/HWM_FS_Jewelry_P2_Recommendations.pdf (accessed 23 October 2008).

932. Clark, P. 1990. *Waste Treatment and Metal Recovery at the Robbins Company* (PB-95–237269). Boston: Massachusetts Department of Environmental Management.

933. Grant, L. J. 1998. Regulations and Safety in Medical Equipment Design. *Anaesthsia* 53(1): 1–3.

934. Prates, G. A., et al. *Environmental Management Performance in a Dental Manufacturing Company*. Sao Paulo, Brazil: University of Ribeiro.

935. Sedgewick, D. 1971. Solving the Problem of Dust Extraction in the Dental Laboratory. *Dental Technology* 30(5): 10–11.

READINGS IN HEALTH AND SAFETY

936. Anonymous. 1997. Trends in Medical Equipment Service. *Health Devices* 26(6): 246–253.

937. Autian, J. 1972. The Potential Toxic Hazards in the Dental Laboratory. *NADL Journal* 19(9): 5.

938. Calnan, C. D. and Connor, B. L. 1972. Carbon Paper Dermatitis Due to Nigrosine. *Berufsdermatosen* 20(5): 248–254.

939. Dubrow, R. and Gute, D. M. 1987. Cause-specific Mortality among Rhode Island Jewelry Workers. *American Journal of Industrial Medicine* 12(5): 579–593.

940. Feigal, D. W., et al. 2003. Ensuring Safe and Effective Medical Devices. *NEJM* 348(3): 191–192.

941. Foster, C. D., et al. 1988. Dental Laboratory Vacuum Systems. *Biomedical Science Instruments* 24: 139–143.

942. Grant, L. J. 1998. Regulations and Safety in Medical Equipment Design. *Anaesthsia* 53(1): 1–3.

943. Hagdrup, H., et al. 1994. Contact Allergy to 2-hydroxy-5-tert-butyl Benzyl-alcohol and 2,6-bis(hydroxymethyl)-4-tert-butylphenol, Components of a Phenolic Resin Used in Marking Pens. *Contact Dermatitis* 31(3): 154–156.

944. Jaakkola, M. S. and Jaakkola, J. J. 1999. Office Equipment and Supplies: A Modern Occupational Health Concern? *American Journal of Epidemiology* 150(11): 1223–1228.

945. Jagger, D. C., et al. 1995. Cross-infection Control in Dental Laboratories. *British Dental Journal* 179(3): 93–96.

946. Jain, S. M., et al. 1977. Silicosis in Slate Pencil Workers. *Chest* 71(3): 423–426.

947. Kern, D. G., et al. 1992. Malignant Mesothelioma in the Jewelry Industry. *American Journal of Industrial Medicine* 21(3): 409–416.

948. Kugel, G., et al. 2000. Disinfection and Communication Practices: A Survey of U.S. Dental Laboratories. *Journal of the American Dental Association* 131(6): 786–792.

949. Lee, H. S., et al. 1993. Burnisher's Asthma. *Singapore Medical Journal* 34(6): 565–566.

950. Lee, J. S., et al. 1983. Dental Laboratory Health Hazards. *Dental Laboratory Review* 58(11): 23–26.

951. Lombardi, C., et al. 2003. Allergy and Skin Disease in Musicians. *Allergy & Immunology* 35(2): 52–55.

952. Maki, B. and Laszlo, C. A. 1980. Mechanical Hazards in Clinical Equipment. *Journal of Clinical Engineering* 5(2): 133–138.

953. Marin, M. L., et al. 1998. Analysis of Potentially Toxic Phalate Plasticizers Used in Toy Manufacturing. *Bulletin of Environmental Contamination & Toxicology* 60(1): 68–73.

954. Moller, M., et al. 1983. Mutagenicity of Extracts from Typewriter Ribbons and Related Items. *Mutation Research* 119(3): 239–249.

955. Nelson, E. W., et al. 1996. Magnetic Resonance Imaging Characteristics of a Lead Pencil Foreign Body in the Hand. *Journal of Hand Surgery* 21(1): 100–103.

956. Orfan, N. A., et al. 1994. Occupational Asthma in a Latex Doll Manufacturing Plant. *Journal of Allergy & Clinical Immunology* 94(5): 826–830.

957. Palenik, C. J. 1997. Eye Protection in Dental Laboratories. *Journal of Dental Technology* 14(7): 22–26.

958. Pang, A. 1976. Relation of Musical Wind Instruments to Malocclusion. *Journal of the American Dental Association* 92(3): 565–570.

959. Parkhurst, H. J. 1951. Dermatitis from Nail Lacquer Applied to Costume Jewelry. *AMA Archives of Dermatology* 63(2): 264–265.

960. Pichirallo, J. 1971. Lead Poisoning Risks for Pencil Chewers? *Science* 173(996): 509–510.

961. Rosen, L. 2004. Applying Industrial Engineering Practices to Radiology. *Radiological Management* 26(6): 32–35.

962. Runnells, R. R. 1985. Infection Control in the Dental Laboratory. *Trends & Technology in Contemporary Dental Labs* 2(1): 11–2, 14, 16.

963. Rustemeyer, T. and Frosch, P. J. 1996. Occupational Skin Diseases in Dental Laboratory Technicians. *Contact Dermatitis* 34(2): 125–133.

964. Schplowsky, A. F. 1971. Lead in Paint on Pencils. *HSMHA Health Report* 86(11): 961–962.

965. Trautman, T. 1983. Health and Safety in the Dental Laboratory. *NADL Journal* 30(3): 26–28.

966. Verran, J., et al. 1996. Microbiological Study of Selected Risk Areas in Dental Technology Laboratories. *Journal of Dentistry* 24(1–2): 77–80.

967. Walter, W. G. and Chaffey, D. 1959. Bacteriological and Cleaning Studies on the Mouthpieces of Musical Instruments. *Applied Microbiology* 7(2): 126–130.

968. Wear, J. O. 1999. Technology and the Future of Medical Equipment Maintenance. *Health Estate* 53(4): 12, 14, 18–19.

969. Wittman, R. and Hu, H. 2002. Cadmium Exposure and Nephropathy in a 28-Year-Old Female Metals Workers. *Environmental Health Perspectives* 110(12): 1261–1266.

970. Zadeh, A. N. and Dufresne, A. 1998. Chemical Hazards in Dental Laboratories. *Indoor Built Environment* 7(3): 146–155.

Sector 42: Wholesale Trade

421: Wholesale Trade, Durable Goods
4211: Motor Vehicle and Motor Vehicle Parts and Supplies Merchant Wholesalers
4212: Furniture and Home Furnishing Merchant Wholesalers
4213: Lumber and Other Construction Materials Merchant Wholesalers
4214: Professional and Commercial Equipment and Supplies Merchant Wholesalers
4215: Metal and Mineral (except Petroleum) Merchant Wholesalers
4216: Electrical and Electronic Goods Merchant Wholesalers
4217: Hardware, Plumbing, and Heating Equipment and Supply Wholesalers
4218: Machinery, Equipment, and Supplies Wholesalers
4219: Miscellaneous Durable Goods Merchant Wholesalers

READINGS IN HISTORY

1. Albright, J. 1941. Changes in Wholesaling, 1929–1939. *Journal of Marketing* 6(1): 31–37.

2. Alt, L. 1990. Photographic Equipment and Supplies. In *Manufacturing: A Historiographical and Bibliographical Guide,* ed. D. O. Whitten, 367–382. Westport, CT: Greenwood Press.

3. Barber, C. L. 1973. The Farm Machinery Industry. *American Journal of Agricultural Economics* 55(5): 820–828.

4. Bayfield, W. W. 1949. Wholesale Coal Industry. *Journal of Marketing* 14(2): 352–354.

5. Becker, W. H. 1971. American Wholesale Hardware Trade Associations, 1870–1900. *Business History Review* 45(2): 179–200.

6. Cassels, J. M. 1936. The Marketing Machinery of the United States. *The Quarterly Journal of Economics* 50(4): 658–679.

7. Darling, S. L. 1949. The Lumber Wholesaler. *Journal of Marketing* 14(2): 349–351.

8. Engelke, W. J. 1966. The Distribution of Selected Current-Carrying Wiring Devices through Hardware Wholesalers. Thesis, Northern Illinois University, Dekalb.

9. Friedman, L. M. 1947. The Drummer in Early American Merchandise Distribution. *Bulletin of the Business Historical Society* 21(2): 39–44.

10. Genesove, D. 1993. Adverse Selection in the Wholesale Used Car Market. *The Journal of Political Economy* 101: 644–665.

11. Haring, A. 1940. Wholesaling and the Wholesaler. *Annals of the American Academy of Political and Social Science* 209: 39–45.

12. Hochfelder, D. and Helper, S. 1996. Suppliers and Product Development in the Early American Automobile Industry. *Business & Economic History* 25(2): 39–51.

13. Hovde, H. T. 1949. Wall Paper Wholesaling. *Journal of Marketing* 14(2): 342–345.

14. Hovey, N. H. 1939. Distribution Problems of the Red Cedar Shingle Industry. *Journal of Marketing* 4(2): 157–167.

15. Kelsey, C. B. 1949. Wholesale Furniture Business. *Journal of Marketing* 14(2): 339–341.

16. Leigh, W. W. 1936. Wholesaling of Automobile Tires. *Journal of Marketing* 1(2): 91–105.

17. Lewis, E. H. 1960. The Distribution of Industrial Electrical Products. *Journal of Marketing* 24(4): 23–28.

18. Lord, J. D. 1984. Shifts in the Wholesale Trade Status of U.S. Metropolitan Areas. *The Professional Geographer* 36(1): 51–63.

19. Marx, T. G. 1985. The Development of the Franchise Distribution System in the U.S. Automobile Industry. *The Business History Review* 59(3): 465–474.

20. Olney, M. L. 1990. Demand for Consumer Durable Goods in 20th Century America. *Explorations in Economic History* 27(3): 322–349.

21. Parks, R. W. 1974. The Demand and Supply of Durable Goods and Durability. *The American Economic Review* 64(1): 37–55.

22. Pyle, C. G. 1949. Electrical Appliance Wholesaling. *Journal of Marketing* 14(2): 336–338.

23. Rubenstein, J. M. 1988. Changing Distribution of American Motor Vehicle Parts Suppliers. *Geographical Review* 78(3): 288–298.

24. Ruark, B. W. 1949. Automotive Wholesaling. *Journal of Marketing* 14(2): 346–348.

25. Simmons, P. B. 1987. Your New Competitors: Wholesalers. *Hospital Entrepreneur Newsletter* 3(4): 1–7.

26. Tillman, T. Y. 1960. A Study of the Automobile Replacement Parts Industry to Evaluate the Role of the Automotive Parts Wholesaler. Thesis, University of Florida, Gainesville.

27. Vincent, C. E. 1964. The Retail Hardware Decline. *Journal of Marketing* 28(1): 26–29.

28. Weder, M. 1998. Fickle Consumers, Durable Goods, and Business Cycles. *Journal of Economic Theory* 81(1): 37–57.

29. Wood, H. S. 1933. Wholesale Distribution of Sporting Goods. Thesis, Ohio State University.

30. Yost, J. R. 1998. Components of the Past and Vehicles of Change: Parts Manufacturers and Supplier Relations in the U.S. Automobile Industry. Dissertation, Case Western Reserve University.

READINGS IN HAZARDOUS WASTE MANAGEMENT AND POLLUTION PREVENTION

31. Das, C. and Tyagi, R. 1994. Wholesaler: A Decision Support System for Wholesale Procurement and Distribution. *International Journal of Physical Distribution and Logistics Management* 24(10): 4.

32. Rondinelli, D. and London, T. 2003. How Corporations and Environmental Groups Cooperate: Assessing Cross-Sector Alliances and Collaborations. *Executive* 17(1): 61–76.

33. U.S. Environmental Protection Agency. 2004. *A Guide to Waste Reduction at Shopping Centers* (EPA/530/R-04/031). Washington, DC: EPA.

READINGS IN HEALTH AND SAFETY

34. Carson, H. J. and Peter, S. 1999. Four Deaths Due to Carbon Monoxide Poisoning in Car Washes. *American Journal of Forensic Medicine & Pathology* 20(3): 274–276.

35. Mathur, A. K., et al. 1991. Sensitivity to Chromium-Containing Automobile Parts. *Contact Dermatitis* 24(1): 65–66.

36. Mirer, F. E. 2003. Mortality in the Motor Vehicle and Motor Vehicle Parts Industry. *Journal of Occupational and Environmental Medicine* 45(8): 789–792.

37. Ulin, S. S. and Keyserling, W. M. 2004. Case Studies of Ergonomic Interventions in Automotive Parts Distribution Operations. *Journal of Occupational Rehabilitation* 14(4): 307–326.

422: Wholesale Trade, Nondurable Goods
4221: Paper and Paper Product Merchant Wholesalers
4222: Drugs and Druggists' Sundries Merchant Wholesalers
4223: Apparel, Piece Goods, and Notions Merchant Wholesalers
4224: Grocery and Related Product Wholesalers
4225: Farm Product Raw Material Merchant Wholesalers
4226: Chemical and Allied Products Merchant Wholesalers
4227: Petroleum and Petroleum Products Merchant Wholesalers
4228: Beer, Wine, and Distilled Alcoholic Beverage Merchant Wholesalers
4229: Miscellaneous Nondurable Goods Merchant Wholesalers

READINGS IN HISTORY

38. Burkart, A. J. 1964. Recent Trends in Periodical Publishing. *The Journal of Industrial Economics* 13(1): 11–22.

39. Cassady, R. and Jones, W. L. 1949. The Los Angeles Wholesale Grocery Structure: 1920–1946. *Journal of Marketing* 14(2): 169–177.

40. Converse, P. D. 1957. 25 Years of Wholesaling: A Revolution in Food Wholesaling. *Journal of Marketing* 22(1): 40–53.

41. Costa, D. L. 2000. Hours of Work and the Fair Labor Standards Act: A Study of Retail and Wholesale Trade, 1938–1950. *Industrial and Labor Relations Review* 53(4): 648–664.

42. Fenili, R. 1985. The Impact of Decontrol on Gasoline Wholesalers and Retailers. *Contemporary Policy Issues* 3(3 Part 2): 119–130.

43. Gearreald, T. N. 1938. *A Study of Fruit and Vegetable Wholesaling and Jobbing Firms in New York City*. Ithaca, NY: Cornell University Press.

44. Grassel, M. A. 1987. The Wholesaling of Washington Wine. Dissertation, University of Washington.

45. Grether, E. T. 1932. The Wholesale Poultry Business in San Francisco. *Journal of Farm Economics* 14(4): 630–639.

46. Hinson, R. A., et al. 1995. Transaction Methods among U.S. Wholesale Nurseries. *Agribusiness* 11(2): 147–154.

47. Kitchen, C. W. 1949. Fresh Fruit and Vegetable Distribution. *Journal of Marketing* 14(2): 325–329.

48. Miller, L. J. 1998. Merchandising to the Mind: The Cultural and Economic Context of Book Retailing and Wholesaling in the United States. Dissertation, University of California, San Diego.

49. Moulton, H. G. 1922. The Case of the Danner-Kraft Dry Goods Company. *The Journal of Political Economy* 30(4): 567–581.

50. Naert, P. and Swinnen, R. 1977. Regulation and Efficiency in Drug Wholesaling. *The Journal of Industrial Economics* 26(2): 137–149.

51. Newcomb, E. L. 1949. The Wholesale Drug Business. *Journal of Marketing* 14(2): 319–321.

52. Patrey, H. B. 1949. Wholesale Tobacco Trade. *Journal of Marketing* 14(2): 330–332.

53. Roth, H. L. 1976. The Book Wholesaler: His Forms and Services. *Library Trends* 24(4): 673–682.

54. Sielaff, R. O. 1945. Postwar Wholesale Dry Goods Sales and Stores. *Journal of Marketing* 10(1): 60–62.

55. ———. 1950. Wholesale Clothing Markets Centered in Duluth. *Journal of Marketing* 15(1): 64–66.

56. Still, R. R. 1953. Historical and Competitive Aspects of Grocery Wholesaling in Seattle, Washington. Thesis, University of Washington.

57. Stout, T. T. and Hawkins, M. H. 1968. Implications of Changes in the Methods of Wholesaling Meat Products. *American Journal of Agricultural Economics* 50(3): 660–675.

58. Toulme, M. L. 1949. Wholesale Grocery Industry. *Journal of Marketing* 14(2): 322–324.

59. Tracey, M. B. 1949. The Present Status of Frozen Food Marketing. *Journal of Marketing* 13(4): 470–480.

60. Waldorf, W. H. 1966. Labor Productivity in Food Wholesaling and Retailing, 1929–1958. *Review of Economics & Statistics* 48(1): 88–93.

61. Wardell, W. L. 1949. Wholesale Shoe Findings Business. *Journal of Marketing* 14(2): 355–357.

62. Williams, W. F. 1958. Structural Changes in the Meat Wholesaling Industry. *Journal of Farm Economics* 40(2): 315–329.

63. Worling, P. M. 1992. Pharmaceutical Wholesaling—An Historical Perspective. *Pharmacy History* 22(4): 3–7.

READINGS IN HAZARDOUS WASTE MANAGEMENT & POLLUTION PREVENTION

64. California Integrated Waste Management Board. 2002. *Waste Reduction at Retail Stores.* http://www.ciwmb.ca.gov/bizWaste/FactSheets/Retail.htm (accessed 23 October 2008).

65. U.S. Environmental Protection Agency. 2000. *Profile of the Agricultural Livestock Production Industry* (EPA/310/R-00/002). Washington, DC: EPA.

66. ——. 2004. *A Guide to Waste Reduction at Shopping Centers* (EPA/530/R-04/031). Washington, DC: EPA.

67. Woodard F. E., et al. 1972. Abatement of Pollution from a Poultry Processing Plant. *Journal of the Water Pollution Control Federation* 44(10): 1909–15.

Sector 44–45: Retail Trade

441: Motor Vehicle and Parts Dealers
4411: Automobile Dealers
4412: Other Motor Vehicle Dealers
4413: Automotive Parts, Accessories, and Tire Stores

READINGS IN HISTORY

1. Berkovec, J. 1985. New Car Sales and Used Car Stocks: A Model of the Automobile Market. *The RAND Journal of Economics* 16(2): 195–214.
2. Bonnie, D. M. 1975. The Jargon of Car Salesman. *American Speech* 50(3/4): 311–313.
3. Borton, W. M. 1944. Buying for Retail Furniture Stores. Dissertation, Ohio State University.
4. Dankert, C. E. 1931. Trends and Problems in the Marketing of Furniture. *The Journal of Business of the University of Chicago* 4(2): 149–173.
5. Duke, J. 1977. New Car Dealers Experience Long-term Gains in Productivity. *Monthly Labor Review* 100(3): 29–33.
6. Fox, B. J. 1990. Selling the Mechanized Household: 70 Years of Ads in the *Ladies Home Journal*. *Gender & Society* 4(1): 25–40.
7. French, M. 1986. Structural Changes and Competition in the U.S. Tire Industry, 1920–1937. *The Business History Review* 60(1): 28–54.

8. Leonard, W. N. and Weber, M. G. 1970. Automakers and Dealers: A Study of Criminogenic Market Forces. *Law & Society Review* 4(3): 407–424.

9. O'Brien, A. P. 1997. The Importance ofAdjusting Production to Sales in the Early Automobile Industry. *Explorations in Economic History* 34(2): 195–219.

10. Olney, M. L. 1989. Credit as Production-smoothing Device: The Case of Automobiles, 1913–1938. *Journal of Economic History* 49(2): 377–391.

11. Rush, R. G. 1959. The Franchising of Automobile Dealers Since 1955. Dissertation, Ohio State University.

12. Schalk, A. F. 1948. Significant Merchandising Trends of the Independent Tire Dealer. *Journal of Marketing* 12(4): 462–469.

13. Vincent, C. E. 1964. The Retail Hardware Decline. *Journal of Marketing* 28(1): 26–29.

READINGS IN HAZARDOUS WASTE MANAGEMENT AND POLLUTION PREVENTION

14. Calta, M. and Kleiner, B. H. 2001. Managing Automobile Dealerships Effectively. *Management Research News* 24 (3/4): 22–28.

15. Florida Department of Environmental Protection. 2004. *Florida Green Yards: An Environmental Compliance Workbook for Automotive Recyclers.* http://www.dep.state.fl.us/central/Home/Green_Yards/Compliance_Wkb_SalvageYards.pdf (accessed 23 October 2008).

16. Longworth, J. 2003. Green Concerns at Car Dealerships: A Competitor Analysis. Master's thesis, Lulea University of Technology.

17. Nieuwenhuis, P. and Wells, P. E. 2003. *The Automotive Industry and the Environment: A Technical, Business and Social Future*. Boca Raton, FL: CRC Press.

18. Washington State Department of Ecology. 1995. *Best Management Practices for Automobile Dealerships—Auto Wastes & Containers* (95–405A). Olympia, WA: Washington State Department of Ecology.

READINGS IN HEALTH AND SAFETY

19. Aziz, A. 2005. Relationship between Machiavellianism Scores and Performance of Automobile Salespersons. *Psychological Reports* 94(2): 464–466.

20. Paulozzi, L. J. 2005. The Role of Sale of New Motorcycles in a Recent Increase in Motorcycle Mortality Rates. *Journal of Safety Research* 36(4): 361–364.

21. Ytterhus, B. E., et al. 1999. Environmental Initiatives in the Retailing Sector: An Analysis of Supply Chain Pressures and Partnerships. *Ecomanagement and Auditing* 6(4): 181–188.

442: Furniture and Home Furnishings Stores

READINGS IN HISTORY

22. Barnett, K. S. 2006. Furniture Music: The Phonograph as Furniture, 1900–1930. *Journal of Popular Music Studies* 18(3): 301.

23. Dankirt, C. E. 1931. The Marketing of Furniture. *The Journal of Business of the University of Chicago* 4(1): 26–47.

24. Davis, H. S. 1943. Vertical or Horizontal? *Textile Research Journal* 13(14): 30–35.

25. Gibbs, J. 1994. *Curtains and Draperies: History, Design and Inspiration.* Woodstock, NY: Overlook Press.

26. Keeble, T. Turning Houses into Homes. A History of the Retailing and Consumption of Domestic Furnishings. *Journal of Design History* 19(1): 98–100.

27. Lee, S. Fashion Cycles in Floor Coverings, 1950–2000. Dissertation, Auburn University, Alabama.

28. Little, N. F. 1975. Floor Coverings. *American Art Journal* 7(1): 107–117.

READINGS IN HAZARDOUS WASTE MANAGEMENT AND POLLUTION PREVENTION

29. U.S. Environmental Protection Agency. 2004. *A Guide to Waste Reduction at Shopping Centers* (EPA/530/R-04/031). Washington, DC: EPA.

443: Electronics and Appliance Stores

READINGS IN HISTORY

30. Bowden, S. and Offer, A. 1994. Household Appliances and the Use of Time: The U.S. and Britain since the 1920s. *Economic History Review* 47(4): 725–748.

31. Burbrink, P. W., et al. 1972. *Selling and Management in Appliance Retailing*. Austin: University of Texas.

32. Campbell-Kelly, M. 1995. Development and Structure of the International Software Industry, 1950–1990. *Business & Economic History* 24(2): 73–110.

33. ——. 2001. Not Only Microsoft: The Maturing of the Personal Computer Software Industry, 1982–1995. *Business History Review* 75(1): 103–145.

34. Kipnis, S. and Huffstutler, C. 1990. Productivity Trends in the Photographic Equipment and Supplies Industry. *Monthly Labor Review* 113(6): 39–49.

35. Lifshey, E. 1973. *The Housewares Story: A History of the American Housewares Industry*. Chicago: National Housewares Manufacturers Association.

36. Lowrey, J. R. 1954. Distribution Policies and Practices in the Field of Major Electrical Appliances. Dissertation, Ohio State University, Columbus.

37. Nickles, H. 1999. America at Home: A Celebration of 20th Century Housewares. *Technology and Culture* 40(1): 180–182.

38. Rada, E. L. 1957. Financing the Flow of Major Home Appliances. *The Journal of Finance* 12(4): 494–495.

39. Wright, W. 1931. Appliance Merchandising by Public Utilities. *Journal of Land & Public Utility Economics* 7(4): 386–393.

READINGS IN HAZARDOUS WASTE MANAGEMENT AND POLLUTION PREVENTION

40. U.S. Environmental Protection Agency. 2004. *A Guide to Waste Reduction at Shopping Centers* (EPA/530/R-04/031). Washington, DC: EPA.

444: Building Material and Garden Equipment and Supply Dealers

READINGS IN HISTORY

42. Barton, S. S. 2002. *Establishing and Operating a Garden Center: Requirements and Costs*. Ithaca, NY: Cooperative Extension Service.

43. Burns, D. R. 1995. Garden Centers: Integrating Garden Design Principles with Marketing Strategies. Dissertation, Morgan State University.

44. Ettlinger, S. 1998. *The Complete Illustrated Guide to Everything Sold in Hardware Stores*. New York: Macmillan.

45. Greene, J. W. 2005. Sources for the History of the Building and Construction Industry. *Labor History* 46(4): 495–511.

46. Rose, G. S. 1987. Reconstructing a Retail Trade Area: Tucker's General Store, 1850–1860. *The Professional Geographer* 39(1): 33.

47. Roush, C. 1999. *Inside Home Depot: How One Company Revolutionized an Industry through the Relentless Pursuit of Growth*. New York: McGraw-Hill Professional.

48. Safley, C. D. and Wohlgenant, M. K. 1996. Factors That Influence Consumer Selection of Garden Centers. *Journal of Agribusiness* 13(1): 33–50.

49. Sayed-Saleh, F. A-M. 1966. A Study of Independent Hardware Retailers in Shopping Centers. Dissertation, Ohio State University.

50. Vincent, C. E. 1964. The Retail Hardware Decline. *Journal of Marketing* 28(1): 26–29.

51. Voordijk, H. 2000. The Changing Logistical System of the Building Materials Supply Chain. *International Journal of Operations & Production Management* 20(7): 823–841.

52. Wilder, P. S. and Klarquist, V. L. 1988. Retail Hardware Stores Register Productivity Gains. *Monthly Labor Review* 111: 39–40.

READINGS IN HAZARDOUS WASTE MANAGEMENT AND POLLUTION PREVENTION

53. U.S. Environmental Protection Agency. 2004. *A Guide to Waste Reduction at Shopping Centers* (EPA/530/R-04/031). Washington, DC: EPA.

READINGS IN HEALTH AND SAFETY

54. Calvert, G. M., et al. 2007. Acute Poisoning in the U.S. Retail Industry, 1998–2004. *Public Health Reports* 122(2): 232–244.

55. Kraus, J. F., et al. Epidemiology of Acute Low Back Injury in Employees of a Large Home Improvement Retail Company. *American Journal of Epidemiology* 146(8): 637–645.

445: Food and Beverage Stores
4451: Grocery Stores
4452: Specialty Food Stores
4453: Beer, Wine, and Liquor Stores

READINGS IN HISTORY

56. Cox, C. 1994. *Storefront Revolution: Food Co-ops and the Counterculture.* New Brunswick, NJ: Rutgers University Press.

57. Daicoff, D. W. and Houston, D. C. 1985. *The Kansas Retail Liquor Industry.* Lawrence: Institute for Public Policy and Business Research, University of Kansas.

58. Edsall, T. B. 1974. Racetracks, Banks and Liquor Stores. *Washington Monthly* 6(10): 49–55.

59. Freeman, M. 1992. Clarence Saunders: The Piggly Wiggly Man. *Tennessee Historical Quarterly* 51(3): 161–169.

60. Gripsrud, G. and Gronhaug, K. 1985. Structure and Strategy in Grocery Retailing: A Sociometric Approach. *The Journal of Industrial Economics* 33(3): 339–347.

61. Grosberg, M. 1985. Charles Grosberg: Super Market Pioneer. *Michigan Jewish History* 25(1–2): 6–9.

62. Halbrendt, C., et al. 1995. Marketing Problems and Opportunities in Mid-Atlantic Seafood Retailing. *American Journal of Agricultural Economics* 77(5): 1313–1318.

63. Hall, L., et al. 1979. Beef Wholesale–Retail Marketing Margins and Concentration. *Economica* 46(183): 295–300.

64. Hattern, M. I. 1988. Who Really Built the First Supermarket? *Californians* 6(3): 34–37.

65. Helfand, W. H. 1994. The Design of American Pharmacies, 1865–1885. *Pharmacy in History* 36(1): 26–37.

66. Kieckhefer, E. W. 1993. Milwaukee Neighborhood Grocery Stores: A Memoir. *Milwaukee History* 16(2): 34–44.

67. Kuo, R. J., et al. 2002. A Decision Support System for Selecting Convenience Store Location through Integration of Fuzzy AHP and Artificial Neural Network. *Computers in Industry* 47(2): 199–214.

68. Lee, Y. and Koutsopoulos, K. 1976. A Locational Analysis of Convenience Food Stores in Metropolitan Denver. *The Annals of Regional Science* 10(1): 104–117.

69. Lentz, C. M. 1989. Grocery Shopping in the 1930s. *Michigan History* 73(2): 14–15.

70. Lutey, K. 1978. Lutey Brothers Marketeria: America's First Self-service Grocers. *Montana* 28(2): 50–57.

71. MacArthur, J. 1996. The Butcher's Shop: Disgust in Picturesque Aesthetics and Architecture. *Assemblage* 30: 32–43.

72. MacFayden, J. T. 1985. The Rise of the Supermarket. *American Heritage* 36(6): 22–32.

73. Marasco, R. 1974. Food from the Sea: An Economic Perspective of the Seafood Market. *American Journal of Agricultural Economics* 56(5): 1030–1037.

74. Marion, B. W. and Walker, F. E. 1978. Short Run Predictive Models for Retail Meat Sales. *American Journal of Agricultural Economics* 60(4): 667–673.

75. Mayo, J. M. 1993. *The American Grocery Store: The Business Evolution of an Architecture Space*. Westport, CT: Greenwood Press.

76. McCraw, T. K. and Bernstein, J. 1997. *7–Eleven in America and Japan.* Harvard Business School Teaching Note 797–031. Boston: Harvard University.

77. Petroski, H. 2003. The Evolution of the Grocery Bag. *American Scholar* 72(4): 99–111.

78. Schapiro, S. 1998. Bulk Candy Specialty Retailing. *Manufacturing Confectioner* 78(6): 57–60.

79. Seth, A. 2001. *The Grocers: The Rise and Rise of the Supermarket Chains*. London: Kogan Page.

80. Shannon, M. C. 1985. Growth Pattern of Community Pharmacies: Michigan 1877–1973. *Pharmacy in History* 27(4): 192–200.

81. Soper, W. 1983. Supermarkets. *American History Illustrated* 18(1): 40–47.

82. Wagner, J. M. and Berman, O. 1995. Models for Planning Capacity Expansion of Convenience Stores under Uncertain Demand and the Value of Information. *Annals of Operation Research* 59(1): 19–44.

83. Walsh, J. P. 1989. Technological Change and the Division of Labor in Retail Trade: The Supermarket Industry from WW II to the Present. Dissertation, Northwestern University.

84. Whalen, T. J. 1967. State Monopoly of Packaged Liquor Retailing. *Journal of Political Economy* 75: 197.

85. Yee, A. 2003. *Shopping at Giant Foods: Chinese American Supermarkets in Northern California.* Seattle: University of Washington Press.

86. York, J. D. 1987. Retail Liquor Stores Experience Flat Trend in Productivity. *Monthly Labor Review* 110 (2): 25.

READINGS IN HAZARDOUS WASTE MANAGEMENT AND POLLUTION PREVENTION

87. Ledermann, J. and Yount, L. 1992. Quantifying Packaging Waste at Grocery Stores. *Resource Recycling.* (December): 43–45.

88. Lieb, K. 1993. Waste from Grocery Stores. *Waste Age* 24(11): 87–92.

89. Malloy, M. 1991. Grocery Store Recycling. *Waste Age* 22(10): 45–48.

90. Mathur, S. P., et al. 1988. Composting Seafood Wastes. *Biocycle* 29(8): 44–49.

91. Nini, D., et al. 1994. *Creative Solutions for Baker Waste Effluent.* New York: AICE.

92. Purcell, B. and Stetiford, E. 2001. Supermarkets Want Our Sludge: Co-digestion of Supermarket Wastes with Sewage Sludge. *Waste Management* [Vol. #?] 17–19.

93. U.S. Environmental Protection Agency. 2004. *A Guide to Waste Reduction at Shopping Centers* (EPA/530/R-04/031). Washington, DC: EPA.

READINGS IN HEALTH AND SAFETY

94. Chitwood, M. 1970. Nematodes of Medical Significance Found in Market Fish. *American Journal of Tropical Medicine* 19(4): 599–602.

95. Droste, J., et al. 2005. Occupational Exposure among Bakery Workers. *Journal of Occupational & Environmental Medicine* 47(5): 458–465.

96. Faulkner, K. A., et al. 2001. Robbery Characteristics and Employee Injuries in Convenience Stores. *American Journal of Industrial Medicine* 40(6): 703–709.

97. Johnson, E. S. 1994. Cancer Mortality among Workers in the Meat Department of Supermarkets. *Occupational & Environmental Medicine* 51(8): 541–547.

98. Moorland, K., et al. 2002. Neighborhood Characteristics Associated with the Location of Food Stores and Food Service Places. *American Journal of Preventive Medicine* 22(1): 23–29.

99. Osorio, A. M., et al. 1994. Carpal Tunnel Syndrome among Grocery Store Workers. *American Journal of Industrial Medicine* 25(2): 229–245.

100. Ryan, G. A. 1989. The Prevalence of Musculo-skeletal Symptoms in Supermarket Workers. *Ergonomics* 32(4): 359–371.

101. Savage, E. P., et al. 1972. Pesticides Sold in Grocery Stores Are Potential Health Hazards. *Health Services Report* 87(8): 734–736.

102. Worsfold, D. and Griffith, C. J. 2001. An Assessment of Cleaning Regimes and Standards in Butcher Shops. *International Journal of Environmental Health Research* 11(3): 245–256.

446: Health and Personal Care Stores

READINGS IN HISTORY

103. Amlani, A. M. and DeSilva, D. G. 2005. Effects of Economy and FDA Intervention on the Hearing Aid Industry. *American Journal of Audiology* 14: 71–79.

104. Amy-Chin, D. 2001. Sex Offence: The Cultural Politics of Perfume. *Women: A Cultural Review* 12(2): 164–175.

105. Applebaum, P. 1983. Jack A. Robinson and Perry Drug Stores, Inc. *Michigan Jewish History* 23(2): 12–18.

106. Benham, L. 1972. The Effect of Advertising on the Price of Eyeglasses. *Journal of Law & Economics* 15(2): 337–352.

107. Dayenian, M. 1959. Fair Trade and the Retail Pharmacy: A Case Study. Dissertation, University of Wisconsin, Madison.

108. Denzin, N. K., et al. 1968. Incomplete Professionalization: The Case of Pharmacy. *Social Forces* 46(3): 375–381.

109. Helfand, W. H. 1994. The Design of American Pharmacies, 1865–1885. *Pharmacy in History* 36(1): 26–37.

110. Herring, J. P. 1973. The Expanding Role of Chain Drug Stores in Health Care. *American Journal of Pharmacy Science* 145(2): 51–55.

111. Sakely, T. 1981. Nate S. Shapiro and Cunningham Drug Stores, Inc. *Michigan Jewish History* 21(2): 3–15.

112. Savage, D. A. 1994. The Professions in Theory and History: The Case of Pharmacy. *Business & Economic History* 23(2): 129–160.

113. Turner, R. G. 1998. The Hearing Aid Expert: Audiologist, Dealer, or Otolaryngologist? *American Journal of Audiology* 7: 5–20.

114. Zellmer, W. A. 1996. Searching for the Soul of Pharmacy. *American Journal of Health Sys. Pharm.* 53(16): 1911–1916.

READINGS IN HAZARDOUS WASTE MANAGEMENT AND POLLUTION PREVENTION

115. Brown-Soo, P. M. 1991. Environmental Protection Efforts by a Home Care Pharmacy. *American Journal of Hospital Pharmacy* 48(1): 118–119.

116. Campbell, D., et al. 2000. Pharmacy: Waste Not, Want Not. *Health Services Journal* 24 (110): 31.

117. Smith, C. A. 2002. Managing Pharmaceutical Waste: What Pharmacists Should Know. *Journal of the Pharmacy Society of Wisconsin* (November/December): 17–22.

447: Gasoline Stations

READINGS IN HISTORY

118. Bond, E. W. 1982. A Direct Test of the "Lemons" Model: The Market for Used Pickup Trucks. *The American Economic Review* 72(4): 836–840.

119. Chafee, W. 1929. What Is a Super Service Station? *Automobile Digest* 12–13:68–70.

120. Dixon, D. F. 1964. Gasoline Marketing in the United States—The First 50 Years. *Journal of Industrial Economics* 13(1): 23–42.

121. Henderson, W. and Benjamin, S. 1994. *Gas Stations*. Osceola, WI: Motorbooks International.

122. Jakle, J. A. 1978. The American Gasoline Station, 1920–1970. *Journal of American Culture* 1: 520–542.

123. Jakle, J. A. and Sculle, K. A. 1994. *The Gas Station in America*. Baltimore, MD: Johns Hopkins University Press.

124. Jones, C. L. 1922. *Service Station Management: Its Principles and Practice*. New York: Van Nostrand.

125. Lankford, S. 2004. Historic Gas Stations along U.S. 40 in Indiana. Dissertation, Ball State University.

126. Lohof, B. A. 1974. The Service Station in America: The Evolution of a Vernacular Form. *Industrial Architecture* 11: 1–13.

127. Rowcroft, J. A. 1998. *The Truck Stops Here! The History and Physical Development of the Truck Stop Industry in the United States*. Ithaca, NY: Cornell University Press.

128. Schick, J. B. 1977. Vehicular Religion and the Gasoline Service Station. *Midwest Quarterly* 19(1): 104–112.

129. Sculle, K. A. 2000a. Atlantic Refining Company's Monumental Service Stations in Philadelphia, 1917–1919. *Journal of American & Comparative Cultures* 23(2): 39–52.

130. Sculle, K. A. 2000b. Pittsburgh's Monuments to Motoring: Atlantic's Fabulous Stations. *Western Pennsylvania History* 83(3): 122–128.

131. Spellman, S. V. 2004. All the Comforts of Home: The Domestication of the Service Station Industry, 1920–1940. *Journal of Popular Culture* 37(3): 43–477.

132. Tucker, J. M. 1993. Gas Stations: Their Forms, Functions and Evolution through the Century. *Chronicle* 27(1–2): 8–12, 30.

133. Usher, J. M. and Evans, M. G. 1996. Life and Death along Gasoline Alley. *Academy of Management Journal* 39(5): 1428–1466.

134. Vieyra, D. I. 1979. *Fill 'er Up: An Architectural History of America's Gas Stations*. New York: Collier Macmillan.

135. Wheeler, R. C. 1960. Frank Lloyd Wright Filling Station, 1958. *Journal of the Society of Architectural Historians* 19(4): 174–175.

136. Witzel, M. K. 1992. *The American Gas Station*. Osceola, WI: Motorbooks International.

137. ——. 1994. *Gas Station Memories*. Osceola, WI: Motorbooks International.

READINGS IN HAZARDOUS WASTE MANAGEMENT AND POLLUTION PREVENTION

138. Alabama Department of Environmental Management. 1997. *Service Stations & Automotive Repair Shops Waste Reduction Tips*. Montgomery, AL: ADEM Pollution Prevention Unit.

139. Northeast-Midwest Institute. 2002. *Recycling America's Gas Stations*. http://www.nemw.org/recyclegas_stations.pdf (accessed 24 October 2008).

140. Simons, R. A., et al. 1999. The Price and Liquidity Effects of UST Leaks from Gas Stations on Adjacent Contaminated Property. *The Appraisal Journal* (April): 186–194.

READINGS IN HEALTH AND SAFETY

141. Akland, G. G. 1993. Exposure of the General Population to Gasoline. *Environmental Health Perspectives* 101 (Supp). 6: 27–32.

142. Foo, S. C. 1991. Benzene Pollution from Gasoline Usage. *Science of the Total Environment* 103(1): 19–26.

143. Guldberg, P. H. 1992. Gasoline and Vapor Exposures in Service Station and Leaking Underground Storage Tanks Scenarios. *Journal of Exposure Analysis & Environmental Epidemiology* 2(1): 97–107.

144. Hakkola, M. A. and Saarinen, L. H. 2000. Customer Exposure to Gasoline Vapors during Refueling at Service Stations. *Applied Occupational & Environmental Hygiene* 15(9): 677–680.

145. Jo, W. K. and Oh, J. W. 2001. Exposure to Methyl Tertiary Butyl Ether and Benzene in Close Proximity to Service Stations. *JAWMA* 51(8): 1122–1128.

146. Kearney, C. A. and Dunham, D. B. 1986. Gasoline Vapor Exposures at a High Volume Service Station. *AIHA Journal* 47(9): 535–539.

147. Lagorio, S., et al. 1994. Mortality of Filling Station Attendants. *Scandinavian Journal of Work & Environmental Health* 20(5): 331–338.

148. McDermott, H. J. and Vos, G. A. 1979. Service Station Attendants' Exposure to Benzene and Gasoline Vapors. *AIHA Journal* 40(4): 315–321.

149. Page, N. P. and Mehlman, M. 1989. Health Effects of Gasoline Refueling Vapors and Measured Exposures at Service Stations. *Toxicology & Industrial Health* 5(5): 869–890.

150. Periago, J. F., et al. 1997. Evaluation of Environmental Levels of Aromatic Hydrocarbons in Gasoline Service Stations by Gas Chromatography. *Journal of Chromatography* 778(1–2): 263–268.

151. Periago, J. F. and Prado, C. 2005. Evolution of Occupational Exposures to Environmental Levels of Aromatic Hydrocarbons in Service Stations. *Annals of Occupational Hygiene* 49(3): 233–240.

152. Sommer, R. 1991. Gas Station Psychology. *Environment & Behavior* 23(2): 131–149.

153. Vayghani, S. A. and Wesiel, C. 1999. The MTBE Air Concentrations in the Cabin of Automobiles while Fueling. *Journal of Exposure Analysis & Environmental Epidemiology* 9(3): 261–267.

154. Verma, D. K., et al. 2001. Benzene and Total Hydrocarbons Exposures in the Downstream Petroleum Industries. *AIHA Journal* 62(2): 176–194.

155. Wixtrom, R. N. and Brown, S. L. 1992. Individual and Population Exposures to Gasoline. *Journal of Exposure Analysis & Environmental Epidemiology* 2(1): 23–78.

448: Clothing and Clothing Accessories Stores
4481: Clothing Stores
4482: Shoe Stores
4483: Jewelry, Luggage, and Leather Goods Stores

READINGS IN HISTORY

156. Ahmed, A. and Wilder, P. S. 1995. Productivity in Retail Miscellaneous Shopping Goods Stores. *Monthly Labor Review* 118(10): 33.

157. Boxerman, B. A. 1973. The Edison Brothers, Shoe Merchants: Their Georgia Years. *Georgia Historical Quarterly* 57(4): 511–525.

158. Burd, H. A. 1941. Mortality of Men's Apparel Stores in Seattle, 1929–1939. *Journal of Marketing* 6(1): 22–26.

159. Cahen, A. 1949. Measuring the Merchandise Flow of Men's Clothing. *Journal of Marketing* 14(1): 67–71.

160. Carter, C. B. 1909. Hosiery Manufacture in the United States. *Annals of the American Academy of Political & Social Science* 34(3): 101–108.

161. Cook, D. T. 2004. *The Commodification of Childhood: The Children's Clothing Industry and the Rise of the Child Consumer*. Durham, NC: Duke University Press.

162. Derango, J. H. 1974. Strategic Planning and Economic Analysis of Starting a Retail Women's Clothing Shop. Dissertation, United States International University.

163. Fraser, S. 1983. Combined and Uneven Development in the Men's Clothing Industry. *The Business History Review* 57(4): 522–547.

164. Goldman, R. and Papson, S. 1998. *Nike Culture: The Sign of the Swoosh*. Thousand Oaks, CA: Sage Publications.

165. Hall, L. 1975. Neiman-Marcus: The Beginning. *Western States Jewish Historical Quarterly* 7(2): 138–150.

166. Hoffmann, L. J. 1927. Shoe Retailing. Dissertation, University of Cincinnati.

167. Juffer, J. 1996. A Pornographic Femininity? Telling and Selling Victoria's (Dirty) Secrets. *Social Text* 48: 27–48.

168. Preston, C. 1999. What's in Store: The Face of Jewelry Retailing Is Changing. *American Jewelry Manufacturer* (April): 54–61.

169. Spangler, C. D. 1923. The Economic Development of Shoe Retailing in the United States. Dissertation, University of Nebraska.

170. University of Pittsburgh. 1935a. *Some Facts About Negligees and Lounging Pajamas*. Pittsburgh: Research Bureau of Retail Training.

171. ——. 1935b. *Some Facts About Women's Hosiery*. Pittsburgh: Research Bureau for Retail Training.

172. ——. 1936. *Merchandise Facts to Help Sell Handbags and Small Leather Goods*. Pittsburgh: Research Bureau for Retail Training.

173. Zakim, M. 1999. A Ready-made Business: The Birth of the Clothing Industry in America. *The Business History Review* 73(1): 61–90.

READINGS IN HAZARDOUS WASTE MANAGEMENT AND POLLUTION PREVENTION

174. U.S. Environmental Protection Agency. 2004. *A Guide to Waste Reduction at Shopping Centers* (EPA/530/R-04/031). Washington, DC: EPA.

READINGS IN HEALTH AND SAFETY

175. Anonymous. 1953. Radiation Exposure in Shoe Stores Studied by New York Staff. *Occupational Health* 13(5): 76.

176. Arbogast, R. T., et al. 2000. Monitoring Insect Pests in Retail Stores by Trapping and Spatial Analysis. *Journal of Economic Entomology* 93(5): 1531–1542.

177. Byer, T. T. and Morell, D. S. 2004. Periumbilical Allergic Contact Dermatitis: Blue Jeans or Belt Buckles? *Pediatric Dermatology* 21(3): 223–226.

178. Duffin, J. and Hayter, C. R. 2000. Baring the Sole: The Rise and Fall of the Shoe-Fitting Fluoroscope. *Isis* 91(2): 260–282.

179. Gardner, L. I., et al. 1999. Risk Factors for Back Injury in 31,076 Retail Merchandise Store Workers. *American Journal of Epidemiology* 150(8): 825–833.

180. Weidenhamer, J. D. and Clement, M. L. 2007. Widespread Lead Contamination of Imported Low-cost Jewelry in the U.S. *Chemosphere* 67(5): 961–965.

451: Sporting Goods, Hobby, Book, and Music Stores

READINGS IN HISTORY

181. Aduddell, R. M. and Cain, L. P. 1990. Musical Instruments. In *Manufacturing: A Historiographical and Bibliographical Guide,* ed. D. O. Whitten, 385–388. Westport, CT: Greenwood Press.

182. Anonymous. 2005. Book Sense: Independent Bookstores for Independent Minds. *Publishing Research Quarterly* 21(2): 30–34.

183. Artto, K. and Pylkkanen, E. 1999. An Effective Procedure for the Distribution of Magazines. *International Transactions in Operational Research* 6(3): 289.

184. Burke, A. E. 2005. The Demand for Vinyl L.P.s 1975–1988. *Journal of Cultural Economics* 18(1): 41–64.

185. Connolly, M. A. 1995. The Transformation of Home Sewing and the Sewing Machine in America, 1850–1929. Dissertation, University of Delaware.

186. Dillard, M. 1999. The Economics of Electronic Commerce: A Study of Online and Physical Bookstores. Bachelor's honor thesis, University of California, Berkeley.

187. Fuller, B. K. 1974. Retail Store Image as Related to Customer Fabric Buying Practices. Dissertation, University of Delaware, Newark.

188. Garvey, E. G. 1996. *The Adman in the Parlor: Magazines and the Gendering of Consumer Culture, 1880s to 1910s*. New York: Oxford University Press.

189. Godley, A. 2001. The Global Diffusion of the Sewing Machine, 1850–1914. *Research in Economic History* 20: 1–45.

190. Jack, A. B. 1957. The Channels of Distribution for an Innovation: The Sewing Machine Industry in America, 1860–1865. *Explorations in Entrepreneurial History* 9(3): 113–141.

191. Kean, R. C., et al. 1996. Competitive Strategies in the Craft Product Retailing Industry. *Journal of Small Business Management* 34(1): 13–23.

192. Leech, D. S. 1965. A Survey of Specialty Stores Retailing Crafts. Dissertation, University of Tennessee.

193. Ling, S. Y. 1998. The Present and Future of Bookshops. *Journal of Internet Pioneer* 47.

194. Lyon, P. 1958. Isaac Singer and His Wonderful Sewing Machine. *American Heritage* 9(6): 34–39, 103–109.

195. Moe, W. W. and Fader, P. S. Modeling Hedonic Portfolio Products: A Joint Segmentation Analysis of Music Compact Disc Sales. *Journal of Marketing Research* 38(3): 376–385.

196. Nourie, A. and Nourie, B. 1990. *American Mass Market Magazines*. Westport, CT: Greenwood Press.

197. Saanwold, W. 1998. Espresso and Ambiance: What Public Libraries Can Learn from Bookstores. *Library Administration & Management* 12(4): 200–211.

198. Seiter, E. 1992. Toys Are Us: Marketing to Children and Parents. *Cultural Studies* 6(2): 232–247.

199. Thomson, R. 1984. The Eco-technic Process and the Development of the Sewing Machine. *Research in Economic History* (3): 243–269.

200. ——. 1987. Learning by Selling and Invention: The Case of the Sewing Machine. *Journal of Economic History* 47(2): 433–445.

201. Thomson, R. D. 1977. The Origin of Modern Industry in the United States: The Mechanization of Shoe and Sewing Machine Production. Dissertation, Yale University.

READINGS IN HAZARDOUS WASTE MANAGEMENT AND POLLUTION PREVENTION

202. U.S. EPA. 2004. *A Guide to Waste Reduction at Shopping Centers* (EPA/530/R-04/031). Washington, DC: EPA.

READINGS IN HEALTH AND SAFETY

203. Conn, J. M., et al. 2004. Injuries from Paintball Game Related Activities in the U.S., 1997–2001. *Injury Prevention* 10(3): 139–143.

204. Curri, T. B., et al. 2003. Playing with Fire: Images of Fire on Toy Packaging. *Journal of Burn Care Rehabilitation.* 24(3): 163–165.

205. Sanguino, S. M., et al. 2002. Handgun Safety: What Do Consumers Learn from Gun Dealers? *Archives of Pediatric & Adolescent Medicine* 156(8): 777–780.

206. Weinstock, E. and Davis, P. 1985. DMSO, Hobby Shops and the FDA: The Diffusion of a Health Policy Dilemma. *Health Education* 16(3): 23–26.

452: General Merchandise Stores
4521: Department Stores

READINGS IN HISTORY

207. Benson, S. P. 1979. Palace of Consumption and Machine for Selling: The American Department Store, 1880–1940. *Radical History Review* 21: 199–221.

208. ——. 1986. *Counter Culture: Saleswomen, Managers and Customers in American Department Stores, 1890–1940*. Chicago: University of Illinois Press.

209. Davis, E. E., et al. 2006. The Economic and Social Impact of Big-box Retailers. *American Journal of Agricultural Economics* 88(5): 1289.

210. Eskilson, S. 2000. Sears Beautiful. *Chicago History* 29(2): 26–43.

211. Ferry, J. W. 1960. *A History of the Department Store*. New York: Macmillan.

212. Hendrickson, R. 1979. *The Grand Emporiums: The Illustrated History of America's Great Department Stores*. New York: Stein & Day.

213. Hower, R. M. 1946. *History of Macy's New York: 1858–1919*. Cambridge, MA: Harvard University Press.

214. Iarocci, L. 2003. Spaces of Desire: The Department Store in America. Dissertation, Boston University.

215. Klassen, H. C. 1992. T. C. Power & Bros.: The Rise of a Small Western Department Store, 1870–1902. *The Business History Review* 66(4): 671–722.

216. Lancaster, B. 1995. *The Department Store: A Social History*. Leicester, England: Leicester University Press.

217. MacLean, A. M. 1899. Two Weeks in Department Stores. *The American Journal of Sociology* 4(6): 721–741.

218. Munroe, S. 2001. Retail Structure Dynamics and the Forces behind Big-box Retailers. *The Annals of Regional Science* 35(3): 357–373.

219. Resseguie, H. E. 1965. Alexander Turney Stewart and the Development of the Department Store, 1823–1876. *The Business History Review* 39(3): 301–322.

220. Rosenberg, L. J. 1988. *Dillard's: The First Fifty Years*. Fayetteville: University of Arkansas Press.

221. Stone, K. E. 1995. *Competing with the Retail Giants: How to Survive in the New Retail Landscape*. New York: John Wiley & Sons.

222. Weiss, D. A. and Lummis. D. M. 1995. *Value Retailing in the 1990s: Off-Pricers, Factory Outlets and Closeout Stores*. New York: John Wiley & Sons.

READINGS IN HAZARDOUS WASTE MANAGEMENT AND POLLUTION PREVENTION

223. U.S. Environmental Protection Agency. 2004. *A Guide to Waste Reduction at Shopping Centers* (EPA/530/R-04/031). Washington, DC: EPA.

READINGS IN HEALTH AND SAFETY

224. Evans, R. W. 2002. The Sky Is Falling: Head Injuries in Warehouse Retail Stores. *Headache* 42(5): 387–388.

225. Gourdine, S. P., et al. 1983. Health Food Stores Investigation. *Journal of the American Dietetic Association* 83(3): 285–290.

226. Leigh, J. P., et al. 2004. Costs of Occupational Injury and Illness across Industries. *Scandinavian Journal of Work & Environmental Health* 30(3): 199–205.

227. Snellen, J. W. 1962. The Optimal Climate in Department Stores. *British Journal of Industrial Medicine* 19: 165–170.

453: Miscellaneous Store Retailers
4531: Florists
4532: Office Supplies, Stationery, and Gift Stores
4533: Used Merchandise Stores
4539: Other Miscellaneous Store Retailers

READINGS IN HISTORY

228. Boje, D. M. 1991. The Storytelling Organization: A Study of Story Performance in an Office Supply Firm. *Administrative Science Quarterly* 36(1): 106–126.

229. Bradshaw, T. F. 1943. Superior Methods Created the Early Chain Store. *Bulletin of the Business Historical Society* 17(2): 35–43.

230. Foster, G. S. 2004. It All Began with Koppelman: The Jewish Florists of Rhode Island. *Rhode Island Jewish Historical Notes* 14(2): 220–237.

231. Gallanis, P. J. 1999. Pet Supplies: Pure Plays Teach Old Dogs New Tricks. *Discount Store News*. 38 (23): 49.

232. Griner, C. P. 2000. *Floriculture: Designing & Merchandising*. Florence, KY: Thomson Delmar Learning.

233. Gubernick, L. 1996. Marketing: Selling Wax Candles. *Forbes*.

234. Hughes, A. 2000. Retailers, Knowledge and Changing Commodity Networks: The Case of the Cut Flower Trade. *Geoforum* 31(2): 175–190.

235. Jaffe, A. Packaged Sentiments: The Social Meanings of Greeting Cards. *Journal of Material Culture* 4(2): 115–141.

236. Johnson, D. and Johnson, E. 1993. *Selling Antiques and Collectibles: 50 Ways to Improve Your Business*. Radnor, PA: Wallace-Homestead Book Co.

237. Kean, R. C., et al. 1996. Competitive Strategies in the Craft Product Retailing Industry. *Journal of Small Business Management* 34 (1): 13–23.

238. Liebowtiz, S. J. 1982. Durability, Market Structure, and New-Used Goods Models. *The American Economic Review* 72(4): 816–824.

239. MacIrving, R. 1963. A Study of Certain Aspects of Retail Florist Business in Oklahoma. Dissertation, Oklahoma State University.

240. McGrath, M. A. 1989. An Ethnography of a Gift Store: Trappings, Wrappings and Rapture. *Journal of Retailing* 65(4): 421–49.

241. Palmer, C. E. and Forsyth, C. J. 2002. Dealers and Dealing in an Antique Mall. *Sociological Spectrum* 22(2): 171–190.

242. Pfarr, D. R. 1957. The Major Marketing Problems of the Retail Florist. Dissertation, University of Pennsylvania.

243. Plattner, S. 1996. *High Art Down Home: An Economic Ethnography of a Local Art Market*. Chicago: University of Chicago Press.

244. Salt, C. H. 1937. Survey of Selling Methods and Consumer Interest in Flowers. Dissertation, Ohio State University.

245. Sherry, J. F. 1990. A Sociocultural Analysis of a Midwestern American Flea Market. *The Journal of Consumer Research* 17(1): 13–30.

246. Singer, L. 1978. Microeconomics of the Art Market. *Journal of Cultural Economics* 2(1): 21–40.

247. Tyrrell, C. G. 1964. The History and Development of the Winterthur Gardens. *Winterthur Portfolio* 1: 122–138.

248. Wallis, A. D. 1997. *The Rise and Decline of Mobile Homes*. Baltimore, MD: Johns Hopkins University Press.

249. Warren-Boulton, F. R. and Dalkir, S. 2001. Staples and Office Depot: An Event-Probability Case Study. *Review of Industrial Organization* 19(4): 467–479.

READINGS IN HAZARDOUS WASTE MANAGEMENT AND POLLUTION PREVENTION

250. U.S. Environmental Protection Agency. 2004. *A Guide to Waste Reduction at Shopping Centers* (EPA/530/R-04/031). Washington, DC: EPA.

READINGS IN HEALTH AND SAFETY

251. Adams, R. M., et al. 1990. Alstroemeria. A New and Potent Allergen for Florists. *Dermatology Clinics* 8(1): 73–76.

252. Akpinar-Elci, M., et al. 2004. Work-Related Asthma-Like Symptoms among Florists. *Chest* 125(6): 2336–2339.

253. Albert, M. R. 1998. Novelty Shop Itching Powder. *Australasia Journal of Dermatology* 39(3): 188–189.

254. Guin, J. D. and Franks, H. 2001. Fingertip Dermatitis in a Retail Florist's. *Cutis* 67(4): 328–330.

255. Jaakkola, M. S. and Jaakkola, J. J. 1999. Office Equipment and Supplies: A Modern Occupational Health Concern? *American Journal of Epidemiology* 150(11): 1223–1228.

256. Morse, D. L., et al. 1979. Cut Flowers: A Potential Pesticide Hazard. *American Journal of Public Health* 69(1): 53–56.

257. National Association of State Public Health Veterinarians. 2005. Compendium of Measures to Prevent Disease Associated with Animals in Public Settings. *MMWR* 54: 1–12.

258. Pereira, F. 1996. Hand Dermatitis in Florists. *Contact Dermatitis* 34(2): 144–145.

259. Rycroft, R. J. 1996. Dermatitis in Florists. *Seminars in Dermatology* 15(2): 83–86.

260. Schillinger, J. and DuVall, K. S. 2004. Drinking Water Quality and Issues Associated with Water Vending Machines in the City of Los Angeles. *Journal of Environmental Health* 66(6): 25–31, 43.

261. Tenenbaum, D. 2002. Would a Rose Not Smell as Sweet? Problems Stem from the Cut Flower Industry. *Environmental Health Perspectives* 110(5): A240–A247.

454: Nonstore Retailers
4541: Electronic Shopping and Mail-order Houses

4542: Vending Machine Operators
4543: Direct Selling Establishments

READINGS IN HISTORY

262. Anupindi, R., et al. 1998. Estimation of Consumer Demand with Stock-out Based Substitution: An Application to Vending Machine Products. *Marketing Science* 17(4): 406–423.

263. Biggart, N. W. 1989. *Charismatic Capitalism: Direct Selling Organizations in America.* Chicago: University of Chicago Press.

264. Cairns, S. 1996. Delivering Alternatives: Successes and Failures of Home Delivery Services for Food Shopping. *Transport Policy* 3(4): 155–176.

265. Cole, A. H. 1970. The Mystery of Fuel Wood Marketing in the U.S. *Business History Review* 44(3): 339–359.

266. Cox, W. E. 1920. *Cost Accounting for Retail Fuel Dealers*. Seattle: University of Washington.

267. Davis, M. E. 1982. The Effectiveness of In-Home Demonstrations of New Gas Ranges. Dissertation, University of Georgia.

268. Donthu, N. and Gilliland, D. 1996. The Infomercial Shopper. *Journal of Advertising Research* 36 (2): 69–76.

269. Eicoff, A. 1995. *Direct Marketing through Broadcast Media: TV, Radio, Cable, Infomercials, Home Shopping, and More*. Lincolnwood, IL: NTC Business Books.

270. Gehrt, K. C. 1987. Nonstore Retailing: An Empirical Investigation of the Competitive Market Structure and the Influence of Personal and Situational Factors, Dissertation, University of Kentucky.

271. Johnson, R. N. 2006. The Impact of Self-service Bans in the Retail Gasoline Market. *The Review of Economics & Statistics* 82(4): 625–633.

272. Jones, K. and Biasiotto, M. 1999. Internet Retailing: Current Hype or Future Reality? *The International Review of Retail, Distribution and Consumer Research* 9(1): 69–79.

273. Lanyon, A. C. 1939. The Mortality of Baltimore Fuel Dealers. *Journal of Marketing* 4(1): 68–72.

274. Martin, A. 1976. James J. Hill and the First Energy Revolution: A Study in Entrepreneurship, 1865–1878. *The Business History Review* 50(2): 179–197.

275. Michael, S. C. 1994. Competition in Organizational Form: Mail Order Versus Retail Stores, 1910–1940. *Journal of Economic Behavior & Organization* 23(3): 269–286.

276. Palmer, J. W. 1997. Electronic Commerce in Retailing: Differences across Retail Formats. *The Information Society* 13(1): 75–91.

277. Postol, T. A. 1997. Creating the American Newspaper Boy: Middle Class Route Service and Juvenile Salesmanship in the Great Depression. *Journal of Social History* 31(2): 327–345.

278. Salmon, W. J. 1960. *A Study of Selected Retail Fuel Oil Dealers in Greater Boston*. Boston: Harvard University, Graduate School of Business Administration, George F. Baker Foundation.

279. Schreiber, G. R. 1961. *A Concise History of Vending in the USA*. Chicago: Vend.

280. Schwartz, M. B. 1997. *The Retailing Revolution: The Impact of Nonstore Retailing on Shopping Centers*. Washington, DC: Urban Land Institute.

281. Stone, B. and Jacobs, R. 2001. *Successful Direct Marketing Methods*. New York: McGraw-Hill Professional.

282. Vijayasarathy, L. R. and Jones, J. M. 2000. Print and Internet Catalog Shopping: Assessing Attitudes and Intentions. *Internet Research* 10(3): 191–202.

READINGS IN HEALTH AND SAFETY

283. Drescher, M. J., et al. 1999. Heating Oil Company Responses to Inquiries Concerning Carbon Monoxide Toxicity. *Annals of Emergency Medicine* 33(4): 406–408.

284. Hopper, S. H. and Adams, H. S. 1958. Copper Poisoning from Vending Machines. *Public Health Reports* 73(10): 910–914.

285. Koehler, H. H., et al. 1977. Tocopherols in Canned Entrees and Vended Sandwiches. *Journal of the American Dietetic Association* 70(6): 616–620.

286. Lowry, W. T., et al. 1991. Toxicological Investigation of Liquid Petroleum Gas Explosion. *Journal of Forensics Science* 36(2): 386–396.

287. McSwane, D. Z, et al. 1994. Drinking Water Quality Concerns and Water Vending Machines. *Journal of Environmental Health* 56(10): 7–12.

288. Ors, S. 2005. Cold Injury from Liquid Petroleum Gas. *Scandinavian Journal of Plastic & Reconstructive Surgery & Hand Surgery* 39(3): 188–190.

289. Schillinger, J. and Du Vall Knorr, S. 2004. Drinking Water Quality Concerns Associated with Water Vending Machines in the City of Los Angeles. *Journal of Environmental Health* 66(6): 25–31.

Sector 48–49: Transportation and Warehousing

481: Air Transportation
4811: Scheduled Air Transportation
4812: Nonscheduled Air Transportation

READINGS IN HISTORY

1. Baur, E. N. 1963. Problems Maintaining Scheduled Helicopter Operations in the New York Area. *Annals of the New York Academy of Sciences* 107(1): 92–101.
2. Bender, A. R. 1986. Flying on the Cheap: The Morphogenesis of Affordable Airline Transportation in America. Dissertation, University of California Berkeley.
3. Bilstein, R. E. 1967. Putting Aircraft to Work: The First Air Freight. *Ohio History* 76(4).
4. Bishop, J. C. 1992. A History of the U.S. Air Cargo Industry, 1955–1978. *Essays in Economic & Business History* 10: 281–286.
5. Bryan, D. L. and O'Kelly, M. E. 1999. Hub-and-Spoke Networks in Air Transportation: An Analytical Review. *Journal of Regional Science* 39(2): 275–295.
6. Croix, S. J., et al. 1986. Airport Taxi Service Regulation: An Analysis of an Exclusive Contract. *Transportation* 13(2): 145–161.
7. Edwards, E. L. 1980. *An Overview of Commuter Air Carriers and Service to Small Communities*. Berkeley: Institute of Transportation Studies, University of California.
8. Fruhan, W. E. 1972. *The Fight for Competitive Advantage: A Study of the U.S. Domestic Air Trunk Carriers*. Boston: Harvard University.

9. Goetz, A. R. 1992. Air Transportation and Growth in the U.S. Urban System, 1950–1987. *Growth & Change* 23(2): 217–238.

10. Gonenc, R. and Nicoletti, G. 2000. Regulation, Market Structure and Performance in Air Passenger Transportation. *OECD Economic Studies* 32: 183–228.

11. Hall, C. L. 1986. The Impact of Technology on the Development of Regional Air Carriers in the U.S., 1945–1978. Dissertation, Southern Illinois University, Carbondale.

12. Klein, M. 1984. Pullmans in the Sky. *American History Illustrated* 19(4): 38–45.

13. Landry, M. and Ozment, J. 2001. Railroad Renaissance: The Post-1970 Short Line Movement. *Essays in Economic & Business History* 19: 151–162.

14. Leonard, W. N. 1947. Some Problems of Postwar Air Transportation. *The American Economic Review* 37(2): 462–477.

15. Levin, R. C. 1978. Allocation in Surface Freight Transportation: Does Rate Regulation Matter? *The Bell Journal of Economics* 9(1): 18–45.

16. Lobo, I. and Zairi, M. 1999. Competitive Benchmarking in the Air Cargo Industry. *Benchmarking: An International Journal* 6(2): 164–191.

17. MacDonald, L. S. 1968. *Cost and Operational Problems of a Scheduled Helicopter Airline.* Technical Paper 680285. Warrenton, PA: SAE International.

18. Mmeints, G. 1983. Michigan and Its Railroads: A Prologue. *Chronicle* 19(2): 4–7.

19. Quinn, J. J. and Engbaum, R. 1969. *Commuter Air Carriers Concepts, Problems and Potential.* Technical Paper 690649. Warrendale, PA: SAE International.

20. Ronen, D. 2000. Scheduling Charter Aircraft. *The Journal of the Operational Research Society* 51(3): 258–262.

21. Schary, P. B. 1967. The CAB and the All-Cargo Airlines: The Early Years. *Business History Review* 41(3): 272–284.

22. Seely, B. E. 1994. Back to the Future? High-speed Rail and Historical Patterns of American Transportation Development. *Railroad History* 170: 5–14.

23. Stephneson, F. J. and Bender, A. R. 1996. The Future of U.S. Business Air Travel. *Transportation Journal* 35(3): 14–32.

24. Taafee, E. J. 1956. Air Transportation and United States Urban Distribution. *Geographical Review* 46(2): 219–238.

READINGS IN HAZARDOUS WASTE MANAGEMENT AND POLLUTION PREVENTION

25. U.S. Environmental Protection Agency. 1998. *Air Transportation Industry* (EPA/310/R-97/001). Washington, DC: EPA.

26. ——. 2000. *Environmental Screening Checklist and Workbook for Airports and Tenant Operations* (EPA/305/B-00/003). Washington, DC: Office of Enforcement & Compliance Assurance.

READINGS IN HEALTH AND SAFETY

27. Anonymous. 1982. Mortality among Infant Passengers on Airlines. *American Journal of Public Health* 72(5): 497–498.

28. Billings, J. E. and Paul, M. F. 1973. Commercial Airlines Industry: Some Lessons for Health Service Planners. *Medical Care* 11(2): 145–155.

29. Colville, R. N., et al. 2001. The Transport Sector as a Source of Air Pollution. *Atmospheric Environment* 35(9): 1537–1565.

30. International Finance Corporation. 1998. *Environmental, Health and Safety Guidelines for Airports.* http://www.ifc.org/ifcext/sustainability.nsf/AttachmentsByTitle/gui_EHSGuidelines2007_Airports/$FILE/Final+-+Airports.pdf (accessed 24 October 2008).

31. Maurino, D. E. 2000. Human Factors and Aviation Safety: What the Industry Has, What the Industry Needs. *Ergonomics* 43(7): 952–959.

32. Muller, M. 2004. Safety Lessons Taken from Airlines. *British Journal of Surgery* 91(4): 393–394.

33. Rayman, R. B. 2000. Airlines' Emergency Medical Kits. *Aviation Space & Environmental Medicine* 71(11): 1151–1152.

34. Roggla, G., et al. 1999. Seat Space on Airlines. *Lancet* 353(9163): 1532.

35. Thibeault, C., et al. 2002. Emergency Medical Kit for Commercial Airlines. *Aviation Space & Environmental Medicine* 73(6): 612–613.

36. Zelnick, S. D., et al. 2002. Prevention of Carbon Monoxide Exposure in General and Recreational Aviation. *Aviation Space & Environmental Medicine* 73(8): 812–816.

482: Rail Transportation

READINGS IN HISTORY

37. Allen, W. B., et al. 2002. Regional and Short Line Railroads in the United States. *Transportation Quarterly* 56(4): 77–113.

38. Babcock, M. W. 1993. *Short Line Railroads Performance. Transportation Quarterly* 49: (2): 73–86.

39. Bitzan, J., et al. 2003. *The Importance of Short Line Railroads to Rural and Agricultural America.* Fargo, ND: North Dakota State University, Upper Great Plains Transportation Institute.

40. Chaney, M. P. 1986. United Aircraft's Turbotrain: The Transfer of Aerospace Technology to Rail Transportation. *Railroad History* 154: 114–124.

41. Daniels, R. 1997. *Trains across the Continent: North American Railroad History.* South Sioux City, NE: Pacific City Graphics.

42. Fair, M. L. 1930. The ICC and the Railroad Terminal Problem. *Quarterly Journal of Economics* 44(3): 462–492.

43. Friedman, D. D. 1979. In Defense of Long-haul/Short-haul Discrimination. *The Bell Journal of Economics* 10(2): 706–708.

44. Lewis, E. A. 1996. *American Short Line Railway Guide.* 5th ed. Waukesha, WI: Kaimbach Publishing.

45. Martens, J. 1999. Heavy Axle Loads: Short Line Railroads. Master's thesis, North Dakota State University.

46. Mayer, H. M. 1981. Current Trends in Great Lakes Shipping. *Geo Journal* 2(2): 117–122.

47. McIllwraith, T. F. 1976. Freight Capacity and Utilization of the Erie and Great Lake Canals before 1850. *The Journal of Economic History* 36(4): 852–877.

48. Means, N. 2004. Sugarcane, Cotton Fields, and High Water: Building the Louisiana Branch of the Texas & Pacific Railroad. *Louisiana History* 45(4): 445–461.

49. Meints, G. 1983. Michigan and Its Railroads. *Chronicle* 19(2): 4–7.

50. Miller, S. L. and Cover, V. D. 1950. *Rates of Return: Class I Line Haul Railroads of the U.S. 1921–1948.* Pittsburgh: University of Pittsburgh Press.

51. O'Connell, W. E. 1970. The Development of the Private Railroad Freight Car, 1830–1966. *Business History Review* 44(2): 190–209.

52. Perl, A. and Dunn, J. A. 1997. Reinventing Amtrak: The Politics of Survival. *Journal of Policy Analysis & Management* 16(4): 598–614.

53. Peterson, D. L. 1968. The Economic Effects of Technological Innovations on Class I Line Haul Railroads, 1947–1963. Dissertation, University of Pittsburgh.

54. Post, R. C. 1977. Manuscript Sources for Railroad History. *Railroad History* 137: 38–63.

55. Schiffman, D. A. 2003. Shattered Rails, Ruined Credit: Financial Fragility and Railroad Operations in the Great Depression. *The Journal of Economic History* 63: 802–825.

56. White, J. H. 1973. The Railroad Reaches California: Men, Machines, and Cultural Migration. *California Historical Quarterly* 52(2): 131–144.

READINGS IN HAZARDOUS WASTE MANAGEMENT AND POLLUTION PREVENTION

57. Barkan, C. P. L. 2004. Cost Effectiveness of Railroad Fuel Spill Prevention Using a New Locomotive Refueling System. *Transportation Research Part D* 9: 251–262.

58. Brownlee, R. C. 1984. *Railway Industry Wastewater Survey*. Washington, DC: Association of American Railroads.

59. U.S. Environmental Protection Agency. 1996. *Environmental Compliance Handbook for Short Line Railroads*. Washington, DC: Hazardous Waste Ombudsman.

60. ——. 1997. *Profile of the Ground Transportation Industry: Trucking, Railroad and Pipeline* (EPA/310/R-97/002). Washington, DC: EPA.

61. ——. 2000. *Environmental Screening Checklist and Workbook for Short Line Railroads* (EPA/305/B-00/002). Washington, DC: Office of Enforcement & Compliance Assurance.

READINGS IN HEALTH AND SAFETY

62. Colville, R. N., et al. 2001. The Transport Sector as a Source of Air Pollution. *Atmospheric Environment* 35(9): 1537–1565.

63. Hunter, S. J., et al. 1998. Predicting Return to Work. A Long-term Follow-up Study of Railroad Workers after Low Back Injuries. *Spine* 23(21): 2319–2328.

64. Kaplan, I. and Zeligman, I. 1962. Occupational Dermatitis of Railroad Workers. *Archives of Dermatology* 85: 95–102.

65. Landon, P., et al. 2005. Noise Exposures of Rail Workers at a North American Chemical Facility. *American Journal of Industrial Medicine* 47(4): 364–369.

66. Sorainen, E. and Rytkonen, E. 1999. Whole Body Vibration of Locomotive Engineers. *AIHA Journal* 60(3): 409–411.

67. Symonds, R. L. 1994. Psychiatric and Preventative Aspects of Rail Fatalities. *Social Science Medicine* 38(3): 431–435.

68. Wenzl, T. B. 1997. Estimating Magnetic Field Exposures of Rail Maintenance Workers. *AIHA Journal* 58(9): 667–671.

69. Woskie, S. R., et al. 1988. Estimation of the Diesel Exhaust Exposures of Railroad Workers. *American Journal of Industrial Medicine* 13(3): 381–394.

483: Water Transportation
4831: Deep Sea, Coastal, and Great Lakes Water Transportation
4832: Inland Water Transportation

READINGS IN HISTORY

70. Fellin, L., et al. 2001. Measuring Benefits from Inland Waterway Navigation Improvements. *Transportation Quarterly* 55(2): 113–136.

71. Fletcher, D. O. 1962. The Decline of the Great Lakes Package-freight Carriers. *Business History Review* 36(4): 387–407.

72. Haites, E. F., et al. 1975. *Western River Transportation: The Era of Early Internal Development, 1810–1860.* Baltimore, MD: Johns Hopkins University Press.

73. Hilton, G. W. 2002. *Lake Michigan Passenger Steamers*. Stanford, CA: Stanford University Press.

74. Hong, J. T. and Plott, C. R. 1982. Rate Filing Policies for Inland Water Transportation. *The Bell Journal of Economics* 13(1): 1–19.

75. Laurent, J. K. 1993. Trade Associations and Competition in Great Lakes Shipping: The Pre–World War I Years. *Anchor News* 24(3): 44–57.

76. ——. 2002. And Cutthroat Competition Prevented: Concentration and Control in Great Lakes Transportation, 1915–1940. *International Journal of Maritime History* 14(2): 43–84.

77. Meyers, A. D. 1999. Brazos Canal: Early Intracoastal Navigation in Texas. *Southwestern Historical Quarterly* 103(2): 174–189.

78. Micketti, G. F. 1979. The Bradley Transportation Line. *Inland Seas* 35(1): 13–20.

79. Miller, T. L. 1969. Texas Land Grants to Promote Water Transportation. *Texana* 7(2): 136–145.

80. Millet, D. J. 1974. The Saga of Water Transportation into Southwest Louisiana to 1900. *Louisiana History* 15(4): 339–347.

81. Schneiders, R. K. 1997. Dams across the Wide Missouri: Water Transportation, the Corps of Engineers, and Environmental Change along the Missouri Valley, 1803–1993. Dissertation, Iowa State University.

82. Slack, B. 1990. Intermodal Transportation in North America and the Development of Inland Load Centers. *The Professional Geographer* 42(1): 721

83. Sykora, T. A. 1972. 1972: A New Era in Great Lakes Transportation. *Inland Seas* 28(2): 131–133.

84. Thayer, W. 1908. Transportation on the Great Lakes. *Annals of the American Academy of Political & Social Science* 31: 126–138.

85. Williamson, S. H. 1977. The Growth of the Great Lakes as a Major Transportation Resource, 1870–1911. *Research in Economic History* 2: 173–248.

READINGS IN HAZARDOUS WASTE MANAGEMENT AND POLLUTION PREVENTION

86. U.S. Environmental Protection Agency. 2000. *Environmental Screening Checklist and Workbook for Water Transportation Industry* (EPA/305/B-00/004). Washington, DC: EPA.

READINGS IN HEALTH AND SAFETY

87. Colville, R. N., et al. 2001. The Transport Sector as a Source of Air Pollution. *Atmospheric Environment* 35(9): 1537–1565.

484: Truck Transportation
4841: General Freight Trucking
4842: Specialized Freight Trucking

READINGS IN HISTORY

88. Annable, J. E. 1973. The ICC, the IBT and the Cartelization of the American Trucking Industry. *Quarterly Review of Economics & Business* 13(2): 33–48.

89. Grandin, T. 1981. *Livestock Trucking Guide*. South St. Paul, MN: Livestock Conservation Institute.

90. Herbst, A. F. and Wu, J. S. K. 1973. Some Evidence of Subsidization: The U.S. Trucking Industry, 1900–1920. *The Journal of Economic History* 33(2): 417–433.

91. Hewins, D. C. 1982. Regulation Without Historical Justification: The Case of Household Moving. *Research in Economic History* Supp. 2: 71–92.

92. Hopkins, S. A. 1986. The Freight Motor Carrier Industry. Dissertation, University of Colorado, Boulder.

93. Madar, D. 2000. *Heavy Traffic: Deregulation, Trade and Transformation in North American Trucking*. Vancouver: University of British Columbia Press.

94. McKee, M. B. 1991. A Cultural Study on the Lifestyle of the Long Haul Trucker. Dissertation, Bowling Green State University.

95. Miles, W. 1977. Doing Things for Old Mt. Pleasant: A History of the Transport Truck Company. *Michigan History* 61(1): 53–88.

96. Quellet, L. J. 1994. *Pedal to the Metal: The Work Lives of Truckers*. Philadelphia: Temple University Press.

97. Thomas, J. H. 1979. *The Long Haul: Truckers, Truck Stops and Trucking*. Memphis, TN: Memphis State University Press.

98. Thomsen, F. L. and Frankhanel, W. R. 1932. *Livestock Trucking in Missouri*. Columbia, MO: University of Missouri, College of Agriculture.

99. Warren, J. R. 1988. A Moving Story: The History of Bekins Moving and Storage Company. *Portage* 9(3): 16–22.

100. Warrington, S. T. 1936. An Analysis of the Development of Cooperative Trucking of Livestock in Minnesota. Dissertation, University of Minnesota.

READINGS IN HAZARDOUS WASTE MANAGEMENT AND POLLUTION PREVENTION

101. U.S. Environmental Protection Agency. 1997. *Profile of the Ground Transportation Industry: Trucking, Railroad and Pipeline* (EPA/310/R-97/002). Washington, DC: EPA.

102. ——. 2000a. *Environmental Screening Checklist and Workbook for Trucking Industry* (EPA/305/B-00/005). Washington, DC: Office of Enforcement & Compliance Assurance.

103. ——. 2000b. *RCRA in Focus: Motor Freight and Railroad Transportation* (EPA/530/K-00/003). Washington, DC: Office of Solid Waste & Emergency Response.

READINGS IN HEALTH AND SAFETY

104. Colville, R. N., et al. 2001. The Transport Sector as a Source of Air Pollution. *Atmospheric Environment* 35(9): 1537–1565.

105. Couper, F. J., et al. 2002. Prevalence of Drug Use in Commercial Tractor-trailer Drivers. *Journal of Forensic Science* 47(3): 562–567.

106. Mitler, M., et al. 1997. The Sleep of Long-haul Truck Drivers. *NEJM* 337: 755–762.

107. Morrow, P. C. and Crum, M. R. 2004. Antecedents of Fatigue, Close Calls, and Crashes among Commercial Motor-vehicle Drivers. *Journal of Safety Research* 35(1): 59–69.

108. Reed, D. B. and Cronin, J. S. 2003. Health on the Road: Issues Faced by Female Truck Drivers. *AAOHN Journal* 51(3): 120–125.

109. Smith, T. J., et al. 2006. Overview of Particulate Exposures in the U.S. Trucking Industry. *Journal of Environmental Monitoring* 8: 711–720.

110. Solomon, A. J., et al. 2004. Healthcare and the Long Haul: Long Distance Truck Drivers: A Medically Undeserved Population. *American Journal of Industrial Medicine* 46(5): 463–471.

111. Stoohs, R. A., et al. 1995. Sleep and Sleep-disordered Breathing in Commercial Long-haul Truck Drivers. *Chest* 107(5): 1275–1282.

485: Transit and Ground Passenger Transportation
4851: Urban Transit Systems
4852: Interurban and Rural Bus Transportation
4853: Taxi and Limousine Service
4854: School and Employee Bus Transportation
4855: Charter Bus Industry
4859: Other Transit and Ground Passenger Transportation

READINGS IN HISTORY

112. Anderson, J. E. 1978. *Transit System Theory*. Lexington, MA: Lexington Books.

113. Barrett, P. 1983. *The Automobile and Urban Transit: The Formation of Public Policy in Chicago, 1900–1930*. Philadelphia: Temple University Press.

114. Bell, T. 1993. The Lander-Yellowstone Transportation Co. *Wind River Mountaineer* 9(2): 4–28.

115. Bianco, M. J. 1999. Technological Innovation and the Rise and Fall of Urban Mass Transit. *Journal of Urban History* 25(3): 348–378.

116. Cervero, R. 1984. Light Rail Transit and Urban Development. *Journal of the American Planning Association* 50(2): 133–147.

117. Cheape, C. W. 1980. *Moving the Masses: Urban Public Transit in New York, Boston and Philadelphia, 1880–1912*. Cambridge, MA: Harvard University Press.

118. Cudahy, B. J. 1990. *Cash, Tokens and Transfers: A History of Urban Mass Transit in North America*. New York: Fordham University Press.

119. ———. 1995. *Under the Sidewalks of New York: The Story of the Greatest Subway System in the World*. New York: Fordham University Press.

120. Davis, R. A. 1990. The Taxicab Business in San Francisco: A Geographical Analysis. Dissertation, San Francisco State University.

121. Due, J. F. 1976. Urban Mass Transit Policy: A Review Article. *Quarterly Review of Economics & Business* 16(1): 93–105.

122. Everly, J. S., et al. 1993. A Survey of Transportation Services for Children with Disabilities. *American Journal of Occupational Therapy* 47(9): 804–810.

123. Fairfield, J. D. 1995. Rapid Transit: Automobility and Settlement in Urban America. *Reviews in American History* 23(1): 80–85.

124. Foster, M. S. 1979a. City Planners and Urban Transportation: The American Response, 1900–1940. *Journal of Urban History* 5(3): 365–396.

125. ———. 1979b. The Western Response to Urban Transportation: A Tale of Three Cities, 1900–1945. *Journal of the West* 18(3): 37.

126. Fravel, F. D., et al. 1982. Regulatory Policy and Economies of Scale in the U.S. Intercity Bus Industry. *Transportation* 11(2): 173–187.

127. Gerrard, M. B. 1981. Transit Vehicles: A New Cottage Industry. *New York Affairs* 6(4): 89–96.

128. Gilbert, G. and Samuels, R. E. 1983. *The Taxicab: An Urban Transportation Survivor.* Chapel Hill: University of North Carolina Press.

129. Gordon, P. and Richardson, H. 1989. Notes from the Underground: The Failure of Urban Mass Transit. *Public Interest* 94: 77–86.

130. Hall, A. 1966. Streetcar Transportation in Akron, 1883–1947. Dissertation, University of Akron.

131. Hodges, G. R. 2007. *Taxi!: A Social History of the New York City Cabdriver.* Baltimore, MD: Johns Hopkins University Press.

132. Huttman, J. P. 1993. Problems of Urban Transportation and the Automobile Industry in the U.S.A. *Studia Historiae Oeconomicae* 20: 131–150.

133. Jackson, C. 1984. *Hounds of the Road: A History of the Greyhound Bus Company*. Bowling Green, OH: Bowling Green University Press.

134. Kaplan, W. H. 1950. The Rise of Interurban Motor Bus Transportation in California. Dissertation, Stanford University.

135. Kitch, E., et al. 1971. The Regulation of Taxicabs in Chicago. *Journal of Law & Economics* 14(2): 285–350.

136. Lee, S. M. and Moore, L. J. 1977. Multi-criteria School Busing Models. *Management Science* 23(7): 703–715.

137. Mallach, S. 1979. The Origins of the Decline of Urban Mass Transportation in the U.S., 1890–1930. *Urbanism Past & Present* 8: 1–17.

138. McGuire, R. A. and Cott, T. N. V. 1984. Public versus Private Economic Activity: A New Look at School Bus Transportation. *Public Choice* 43(1): 25–43.

139. Molloy, S. 1996. *Trolley Wars: Streetcar Workers on the Line*. Washington, DC: Smithsonian Institution Press.

140. Owings, N. A. 1970. Mass Transit and the Cities: Mobility and Place in America's Future. *Current History* 59(348): 95–99, 112–113.

141. Ray, D. W. 2005. A History of Streetcar Service in Athens, Georgia and Some Possibilities for Its Reintroduction. Dissertation, University of Georgia.

142. Rhodes, J. 1988. *Intercity Bus Lines of the Southwest: A Photographic History.* College Station: Texas A&M University Press.

143. Rosenbloom, S. 1988. *The Mobility Needs of the Elderly.* Special Report No. 218. Washington, DC: Transportation Research Board.

144. Sawers, L. 1979. American Ground Transportation Reconsidered. *Review of Radical Political Economics* 11(3): 66–69.

145. Schrag, Z. M. 2006. *The Great Society Subway: A History of the Washington Metro*. Baltimore, MD: Johns Hopkins University Press.

146. Smythe, D. W. 1937. An Economic History of Local and Interurban Transportation in the East Bay Cities. Dissertation, University of California, Berkeley.

147. Sorel, E. 1992. Our Checkered Past. *American Heritage* 43(8): 90–97.

148. St. Claire, D. J. 1981. The Motorization and Decline of Urban Public Transit. *Journal of Economic History* 41(3): 579–600.

149. Toman, J. A. and Hays, B. S., eds. 1996. *Cleveland's Transit Vehicles: Equipment and Technology*. Kent, OH: Kent State University Press.

150. Turner, R. 1998. *Special Needs Transportation Handbook*. White Buffalo Press.

151. Volpe, J. A. 1970. Urban Transportation—Today and Tomorrow. *Urban & Social Change Review* 4(1): 2–5.

152. Walsh, M. 1994. Iowa's Bus Queen: Helen M. Schultz and the Red Ball Transportation Co. *Annals of Iowa* 53(4): 329–355.

153. Walsh, M. 2000. *Making Connections: The Long-distance Bus Industry in the USA*. Aldershot, England: Ashgate.

154. Weiner, E. 1992. *Urban Transportation Planning in the U.S.: A Historical Overview*. Washington, DC: Office of the Secretary of Transportation, Office of Economics.

155. Yago, G. 1984. *The Decline of Transit: Urban Transportation in German and U.S. Cities, 1900–1970*. Cambridge: Cambridge University Press.

156. Yang, H. and Wong, S. C. 1998. A Network Model of Urban Taxi Service. *Transportation Research Part B: Methodological* 32(4): 235–246.

READINGS IN HAZARDOUS WASTE MANAGEMENT AND POLLUTION PREVENTION

157. Anonymous. 1995. *Waste Control Practices at Bus Maintenance Facilities*. Washington, DC: Transportation Research Board, National Research Council.

158. Cohen, J. T., et al. 2003. Fuels for Urban Transit Buses: A Cost Effectiveness Analysis. *Environmental Science and Technology* 37(8): 1477–1484.

159. Gwilliam, K., et al. 2004. *Reducing Air Pollution from Urban Transport*. Washington, DC: The World Bank.

160. Schimek, P. 1981. Reducing Particulate Matter and Oxides of Nitrogen from Heavy Duty Vehicles: The Urban Bus Case. *Transportation Research Record* 1641: 39–47.

READINGS IN HEALTH AND SAFETY

161. Anderson, R. 1992. The Back Pain of Bus Drivers: Prevalence in an Urban Area of California. *Spine* 17(12): 1481–1488.

162. Bachar, R., et al. 1999. Injuries Due to Falls in Urban Buses: 100 Consecutive Cases. *Harefuah* 137(1–2): 77–78, 86.

163. Brunet, A., et al. 1998. Lifetime Exposure to Traumatic Events among a Sample of City Bus Drivers. *Psychology Report* 83(3) (Part 2): 1155–1160.

164. Chen, J. C., et al. 2004. Occupational and Personal Factors Associated with Acquired Lumbar Spondylolisthesis of Urban Taxi Drivers. *Occupational & Environmental Medicine* 61(12): 992–998.

165. Colville, R. N., et al. 2001. The Transport Sector as a Source of Air Pollution. *Atmospheric Environment* 35(9): 1537–1565.

166. Davis, R. M. 1989. The Inaccessibility of Seat Belts in Taxicabs. *American Journal of Public Health* 79(3): 330–331.

167. Evans, G. W. and Carrere, S. 1991. Traffic Congestion, Perceived Control, and Psycho-physiological Stress among Urban Bus Drivers. *Journal of Applied Psychology* 76(5): 658–663.

168. Evans, G. W. and Johansson, G. 1998. Urban Bus Driving: An International Arena for the Study of Occupational Health Psychology. *Journal of Occupational Health* Psychology 3(2): 99–108.

169. Figa-Talamanca, I., et al. 1996. Effect of Prolonged Auto Vehicle Driving on Male Reproduction Function: A Study among Taxi Drivers. *American Journal of Industrial Medicine* 30(6): 750–758.

170. Funakoshi, M., et al. 2004. Measurement of Whole Body Vibration in Taxi Drivers. *Journal of Occupational Health* 46(2): 119–124.

171. Gershon, R. R., et al. 2005. Health and Safety Hazards Associated with Subways: A Review. *Journal of Urban Health* 82(1): 10–20.

172. Goolsby, P. V. 1995. School Bus Safety Past and Present. Dissertation, University of Mississippi.

173. Halpern, P., et al. 2005. Non-collision Injuries in Public Buses: A National Survey of a Neglected Problem. *Emergency Medicine* 22(2): 108–110.

174. Jo, W. K. and Yu, C. H. 2001. Public Bus and Taxicab Drivers' Work-time Exposure to Aromatic Volatile Organic Compounds. *Environmental Research* 86(1): 66–72.

175. Johnson, C. J., et al. 1975. Carbon Monoxide in School Buses. *American Journal of Public Health* 65(12): 1327–1329.

176. Komisar, A., et al. 1991. Head and Neck Trauma in Taxicabs: A Growing Urban Problem. *Archives of Otolaryngology Head & Neck Surgery* 117(4): 442–445.

177. Lanni, T. 2003. Fine Urban and Precursor Emissions Control for Diesel Urban Transit Buses. *Environmental Pollution* 123(3): 427–437.

178. Martins, P. J., et al. 2003. Increased Plasma Homocysteine Levels in Shift Working Bus Drivers. *Occupational & Environmental Medicine* 60(9): 662–666.

179. Morlok, E. K., et al. 2004. Boarding and Alighting Injury Experience with Different Station Platform and Car Entryway Designs on U.S. Commuter Railroads. *Accident Analysis & Prevention* 36(2): 261–271.

180. Paradis, G., et al. 1989. Mortality in a Historical Cohort of Bus Drivers. *International Journal of Epidemiology* 18(2): 397–402.

181. Raggatt, P. T. and Morrissey, S. A. 1997. A Field Study of Stress and Fatigue in Long-distance Bus Drivers. *Behavioral Medicine* 23(3): 122–129.

182. Ragland, D. R., et al. 1987. Prevalence of Hypertension in Bus Drivers. *International Journal of Epidemiology* 16(2): 208–214.

183. Reilly, E. A. 1995. School Bus Safety: Issues and Controversy. *Journal of Pediatric Health Care* 9(3): 145–148.

184. Schlosberg, R. 1982. Taxi Driving: A Study of Occupational Tension. Dissertation, City University of New York.

185. Spivey, A. 2002. Bad Grades for School Buses. *Environmental Health Perspectives* 110(8): A453.

186. Walsleben, J. A., et al. 1999. Sleep Habits of Long Island Rail Commuters. *Sleep* 22(6): 728–734.

187. Wang, P. D. and Lin, R. S. 2001. Coronary Heart Disease Risk Factors in Urban Bus Drivers. *Public Health* 115(4): 261–264.

188. Wier, E. 2002. Diesel Exhaust, School Buses and Children's Health. *CMAJ* 167(5): 505.

189. Zimmerman, R. 2005. Mass Transit Infrastructure and Urban Health. *Journal of Urban Health* 82(1): 21–32.

486: Pipeline Transportation
4861: Pipeline Transportation of Crude Oil
4862: Pipeline Transportation of Natural Gas
4869: Other Pipeline Transportation

READINGS IN HISTORY

190. Bagwell, M. V. 1973. Pipeline Transportation in the 1970s. *Journal of the Transportation Engineering Division* 99(1): 5–15.

191. Castaneda, C. J. 1993. *Regulated Enterprises: Natural Gas Pipelines and Northeastern Markets, 1938–1954.* Columbus: Ohio State University Press.

192. Castaneda, C. J. 2004. History Beneath the Surface: Natural Gas Pipelines and the National Historic Preservation Act. *The Public Historian* 26(1): 105–121.

193. Coates, P. 2001. The Trans-Alaska Pipeline 20th Birthday. *The Public Historian* 23(2): 63–86.

194. Coburn, L. L. 1988. Eighty Years of U.S. Petroleum Pipeline Regulation. *Journal of Transport History* 9(2): 149–169.

195. Cramer, C. A. 1968. Interconnection and Peak Responsibility in the Natural Gas Pipeline Industry. *Land Economics* 44(2): 229–234.

196. Eiber, R. J., et al. 1987. Outside Forces Cause Most Natural Gas Pipeline Failures. *Oil & Gas Journal* 85: 11.

197. Elsworth, C. C. 1961. Integration into Crude Oil Transportation in the 1930s. *Business History Review* 35(2): 180–210.

198. Johnson, A. M. 1966. The Early Texas Oil Industry: Pipelines and the Birth of an Integrated Oil Industry, 1901–1911. *Journal of Southern History* 32(4): 516–528.

199. Oney, F., et al. 1994. Evaluation of Pipeline Transportation of Hydrogen and Natural Gas Mixtures. *International Journal of Hydrogen Energy* 19(10): 813–822.

200. Starsmore, R. P. 1990. History of a Wet Gas Transportation Pipeline from Design through to Decommissioning. *Corrosion Prevention and Control* 37(5): 117–120.

201. Telson, M. L. 1986. Policy Issues in Oil and Gas Transportation Regulation. *Contemporary Policy Issues* 4(4): 60–64.

202. Yaghi, B. M. and Al-Bemani, A. 2002. Heavy Crude Oil Viscosity Reduction for Pipeline Construction. *Energy Sources, Part A: Recovery, Utilization & Environmental Effects* 24(2): 93–102.

READINGS IN HAZARDOUS WASTE MANAGEMENT AND POLLUTION PREVENTION

203. Bruton, C. R. 1970. Gas Pipeline Construction: A Proposal for Environmental Protection. *Stanford Law Review* 22(5): 1073–1093.

204. Douligeris, C., et al. 1997. Development of a National Marine Oil Transportation System Model. *Spill Science & Technology Bulletin* 4(2): 113–121.

205. Gossen, L. P. and Velichkina, L. M. 2006. Environmental Problems of the Oil and Gas Industry. *Petroleum Chemistry* 46(2): 67–72.

206. Jo, Y. D. and Ahn, B. J. 2005. A Method of Quantitative Risk Assessment for Transmission Pipeline Carrying Natural Gas. *Journal of Hazardous Materials* 123(1–3): 1–12.

207. Litto, R., et al. 2007. Capturing Fugitive Methane Emissions from Natural Gas Compressor Buildings. *Journal of Environmental Management* 84(3): 347–361.

208. Pelley, J. 2001. Design of Natural Gas Pipeline Questioned. *Environmental Science and Technology* 35(19): 394A–395A.

209. Tansel, B., et al. 2000. Spill Prevention Priority Analysis for Reducing Accidental Release Risks during Pipeline Transport. *Journal of Environmental Systems* 28(4): 319–335.

READINGS IN HEALTH AND SAFETY

210. Colville, R. N., et al. 2001. The Transport Sector as a Source of Air Pollution. *Atmospheric Environment* 35(9): 1537–1565.

211. Cooke, R. and Jager, E. 1998. A Probabilistic Model for the Failure Frequency of Underground Gas Pipelines. *Risk Analysis* 18(4): 511–527.

212. Macintosh, R. 1977. Wrongly Connected Gas Pipelines. *Lancet* 2(8032): 307.

213. Meshkati, N. 2006. Safety and Human Factors Considerations in Control Rooms of Oil and Gas Pipeline Systems. *International Journal of Occupational Safety Ergonomics* 12(1): 79–93.

214. Perry, W. D. and Haynes, K. E. 1993. Prioritizing Regulatory Policy in Pipeline Safety. *Journal of Contingencies and Crisis Management* 1(2): 90.

215. Verma, K., et al. 2000. Benzene and Total Hydrocarbon Exposures in the Upstream Petroleum Oil and Gas Industry. *AIHA Journal* 61(2): 255–263.

487: Scenic and Sightseeing Transportation
4871: Scenic and Sightseeing Transportation, Land
4872: Scenic and Sightseeing Transportation, Water
4879: Scenic and Sightseeing Transportation, Other

READINGS IN HISTORY

216. Aron, C. S. 1999. *Working at Play: A History of Vacations in the U.S.* New York: Oxford University Press.

217. Buckley, R. J. 1975. *A History of Tramways: From Horse to Rapid Transit.* North Pomfret, VT: David & Charles.

218. Christensen, N. A. and McCool, S. F. 1990. *The Economic Significance of Motorcoach Tours in Montana.* Missoula: Institute for Tourism and Recreation Research, School of Forestry, University of Montana.

219. Cocks, C. 2001. *Doing the Town: The Rise of Urban Tourism in the U.S., 1850–1915.* Berkeley: University of California Press.

220. Coughenower, D. D. 1986. *Homer, Alaska Charter Fishing Industry Study.* Fairbanks, AK: Alaska Sea Grant College Program, University of Alaska.

221. DeLyser, D. 2005. *Ramona Memories: Tourism and the Shaping of Southern California.* Minneapolis: University of Minnesota Press.

222. Hudsman, L. E. 1994. Tourism and the American West. *Journal of the West.* 33(3): 67–76.

223. Kuehn, D. M., et al. 2005. *New York's Great Lakes Charter Boat Fishing Industry, 1975–2002.* General Technical Report NE-326. Newtown Square, PA: Northeastern Research Station, Forest Service.

224. Lichtkoppler, F. R. Ohio's Lake Erie Charter Fishing Industry, 1985–1994. *American Fisheries* 22(1): 14–21.

225. Liston, S. M. 1968. When Sightseeing Meant Site-seeking. *Aerospace Historian* 15(1): 39–43.

226. Rothman, H. K., ed. 2003. *The Culture of Tourism, The Tourism of Culture: Selling the Past to the Present in the American Southwest.* Albuquerque: University of New Mexico Press.

227. Sandford, M. R. 1987. Tourism in Harlem: Between Negative Sightseeing and Gentrification. *Journal of American Culture.* 10(2): 99–105.

228. Schneigert, Z. 1996. *Aerial Tramways and Funicular Railways.* Oxford, NY: Pergamon Press.

229. Shaffer, M. S. 2001. *See America First: Tourism and National Identity, 1880–1940*. Washington, DC: Smithsonian Institution Press.

230. Souther, J. M. 2004. Landscapes of Leisure: Building an Urban History of Tourism. *Journal of Urban History* 30(2): 257–265.

231. Wrobel, D. M. and Long, P. T., eds. 2001. *Seeing and Being Seen: Tourism in the American West*. Lawrence: University of Kansas Press.

READINGS IN HAZARDOUS WASTE MANAGEMENT AND POLLUTION PREVENTION

232. Davies, T., et al. 2000. *Environmental Implications for the Tourism Industry*. Washington, DC: Resources for the Future.

233. Goodall, B. 1995. Environmental Auditing: A Tool for Assessing the Environmental Performance of Tourism Firms. *Geographical Journal* 161(1): 29–37.

234. Johnson, D. 2002. Environmentally Sustainable Cruise Tourism: A Reality Check. *Marine Policy* (26): 261–270

235. Newman, V. and Sage, S. 1996. Tourism and the Environment: Out on a Limb and Sawing. *Yale F&ES Bulletin* No. 99: 72–80.

236. Schulkin, A. 2002. Safe Harbors: Crafting an International Solution to Cruise Ship Pollution. *Georgetown International Environmental Law Review* 15(1): 105–132.

237. Transport Canada. 2003. *Pollution Prevention Guidelines for the Operation of Cruise Ships under Canadian Jurisdiction*. Ottawa, ON: Marine Safety Directorate.

READINGS IN HEALTH AND SAFETY

238. Colville, R. N., et al. 2001. The Transport Sector as a Source of Air Pollution. *Atmospheric Environment* 35(9): 1537–1565.

239. Cramer, E. 2003. Diarrheal Disease on Cruise Ships, 1990–2000: The Impact of Environmental Health Programs. *American Journal of Preventive Medicine* 24(3): 227–233.

240. Dahl, E. 2001. Passenger Mortalities Aboard Cruise Ships. *International Maritime Health* 52(1–4): 19–23.

241. Lindquist, L-J., and Bjork, P. 2000. Perceived Safety as an Important Quality Dimension among Senior Tourists. *Tourism Economics* 6(2): 151–158.

242. Minooee, A. and Rickman, L. S. 1999. Infectious Diseases on Cruise Ships. *Clinical & Infectious Disease* 29(4): 737–743.

243. Peake, D. E., et al. 1999. Descriptive Epidemiology of Injury and Illness among Cruise Ship Passengers. *Annals of Emergency Medicine* 33(1): 67–72.

244. Rooney, R. M., et al. 2004. A Review of Outbreaks of Foodborne Disease Associated with Passenger Ships. *Public Health Reports* 119(4): 427–434.

245. Toda, M., et al. 2004. Medical Assessment of the Health Effects of Short Leisure Trips. *Archives of Environmental Health* 59(12): 717–724.

488: Support Activities for Transportation
4881: Air Transportation Support Activities
4882: Rail Transportation Support Activities
4883: Water Transportation Support Activities
4884: Road Transportation Support Activities
4885: Freight Transportation Arrangement
4889: Other Support Activities for Transportation

READINGS IN HISTORY

246. Bard, J. F. 1997. An Analysis of a Rail Car Unloading Area for a Consumer Products Manufacturer. *Journal of the Operational Research Society* 48(9): 873–883.

247. Bednarek, J. R. D. 2001. *America's Airports: Airfield Development, 1918–1947*. College Station: Texas A&M University Press.

248. Benekohal, R., et al. 1999. Truck Delay and Traffic Conflicts around Weigh Stations. *Transportation Research Record* 1653: 52–60.

249. Billings, C. E. 1996. *Aviation Automation: Search for a Human Centered Approach*. New York: Taylor & Francis Group, LLC.

250. Collins, W. E., et al. 1981. The Selection of Air Traffic Control Specialists: History and Review of Contributions by the Civil Aeromedical Institute, 1960–1980. *Aviation Space Environmental Medicine* 52(4): 217–240.

251. DeCastillo, B. and Daganzo, B. 1993. Handling Strategies for Import Containers at Marine Terminals. *Transportation Research Part B: Methodological* 27(2): 151.

252. DeNeufville, R. and Odoni, A. R. 2003. *Airport Systems: Planning, Design and Management*. New York: McGraw-Hill.

253. Douglas, D. G. 1997. The Invention of Airports: A Political, Economic and Technological History of Airports in the U.S., 1919–1939. Dissertation, University of Pennsylvania.

254. Fair, M. L. 1961. Port Authorities in the U.S. *Law & Contemporary Problems* 26(4): 703–714.

255. Golaszewski, R. 2002. Reforming Air Traffic Control: An Assessment from the American Perspective. *Journal of Air Transport Management* 8(1): 3–11.

256. Gopalan, R. and Talluri, K. T. 1998. The Aircraft Maintenance Routing Problem. *Operations Research* 46(2): 260–271.

257. Goranson, U. G. 1998. Fatigue Issues in Aircraft Maintenance and Repairs. *International Journal of Fatigue* 20(6): 413–431.

258. Jacobs, D. 1986. *American Tow Trucks: The Custom Wrecker Scene in the USA*. London: Osprey.

259. Jansson, J. O. 1982. *Port Economics*. Cambridge, MA: MIT Press.

260. Kennet, D. M. 1994. A Structural Model of Aircraft Engine Maintenance. *Journal of Applied Econometrics* 9(4): 351–368.

261. Keith, R. C. 2005. *Baltimore Harbor: A Pictorial History*. Baltimore, MD: Johns Hopkins University Press.

262. Latorella, K. A. and Prabhu, P. V. 2000. A Review of Human Error in Aviation Maintenance and Inspection. *International Journal of Industrial Ergonomics* 26(2): 133–161.

263. Luczak, H., et al. 2005. Virtual Organizations in the Courier, Express and Parcel Industry. In *Virtual Organizations: Systems and Practices,* ed. L. M. Camarinha et al. New York: Springer US.

264. Mackay, W. E. 1999. Is Paper Safer? The Role of Flight Strips in Air Traffic Control. *ACM Transactions on Computer Human Interaction* 6(4): 311–340.

265. Midori, R., et al. 2005. Maritime Liner Shipping and the Stevedoring Industry: Market Structure and Competition Strategies. *Maritime Policy & Management* 32(2): 89–106.

266. Mitchell, K. A. 2003. The Hagerstown Airport. *American Aviation Historical Society* 48(3): 215–229.

267. Mundy, R. A. 1992. Evolution of Ground Transportation Management As a Major Airport Function. *Transportation Research Record* 1373: 17.

268. Nordlund, W. J. 1998. *Silent Skies: The Air Traffic Controller's Strike*. Westport, CT: Praeger.

269. Olson, T. F. 1960. Cargo Preference and the American Merchant Marine. *Law & Contemporary Problems* 25(1): 82–105.

270. Pomerantz, C. and Alley, R. W. 1987. *How Many Trucks Can a Tow Truck Tow?* New York: Random House.

271. Posey, C. A. 1993. Airports Everywhere. *Air & Space* 8(2): 72–81.

272. Punakivi, M. and Saranen, J. 2001. Identifying the Success Factors in E-grocery Home Delivery. *International Journal of Retail & Distribution Management* 29(4): 156–163.

273. Rosenstein, M. 2000. *The Rise of Maritime Containerization in the Port of Oakland, 1950–1970*. New York: New York University, Gallatin School of Individualized Study.

274. Ryan, A. 1999. Newport State Airport: A History. *Newport History* 70(1): 20–33.

275. Sauerbier, C. L. 1956. *Marine Cargo Operations*. New York: Wiley.

276. Stinnett, D. C. 2003. The Trouble with Tow Trucks: Federal Preemption of State Law Claims against Tow Truck Companies. *Journal of Texas Consumer Law* 26–36.

277. Usselman, S. W. 1984. Air Brakes for Freight Trains. *The Business History Review* 58(1): 30–50.

278. Wickens, C. D. 1998. *The Future of Air Traffic Control: Human Operators and Automation*. Washington, DC: National Academies Press.

279. Wood, D. F. 1995. *Wreckers & Tow Trucks*. Osceola, WI: Motorbooks International.

READINGS IN HAZARDOUS WASTE MANAGEMENT AND POLLUTION PREVENTION

280. Graham, A. 2005. Airport Benchmarking: A Review of the Current Situation. *Benchmarking: An International Journal* 12(2): 99–111.

281. Jacalone, D. P. Incorporating Recycling into Airports: Opportunities and Complexities. *BioCycle* http://www.p2pays.org/ref%5C06/05482.pdf (accessed 24 October 2008).

282. Paipai, E. 1999. *Guidelines for Port Environmental Management*. Report SR554. Howberry Park, UK: HR Wallingford.

283. Upham, P. 2001. Environmental Capacity of Aviation: Theoretical Issues and Basic Research Directions. *Journal of Environmental Planning & Management* 44(5): 721–734.

284. Woolridge, C. F., et al. 1999. Environmental Management of Ports and Harbors. *Marine Policy* 23(4): 413–425.

READINGS IN HEALTH AND SAFETY

285. Averty, P., et al. 2004. Mental Workload in Air Traffic Control: An Index Constructed from Field Tests. *Aviation and Space Environmental Medicine* 75(4): 333–341.

286. Collins, W. E., et al. 1981. The Selection of Air Traffic Control Specialists: History and Review of Contributions by the Civil Aeromedical Institute, 1960–1980. *Aviation and Space Environmental Medicine* 52(4): 217–240.

287. Chen, T. J., et al. 1992. Effects of Aircraft Noise on Hearing and Auditory Pathway Function of Airport Employees. *Journal of Occupational Medicine* 34(6): 613–619.

288. Colville, R. N., et al. 2001. The Transport Sector As a Source of Air Pollution. *Atmospheric Environment* 35(9): 1537–1565.

289. Landsbergis, P. 1986. Is Air Traffic Control a Stressful Occupation? *Labor Studies Journal* 11(2): 117–134.

290. Luszczynski, K., et al. 1978. Radiation Exposure of Air Cargo Workers at the St. Louis International Airport. *Health Physics* 35(4): 523–527.

291. Niessen, C., et al. 1999. Modeling Cognitive Processes of Experienced Air Traffic Controllers. *Ergonomics* 42(11): 1507–1520.

292. Passchier, W., et al. 2000. Public Health Impact of Large Airports. *Review of Environmental Health* 15(1–2): 83–96.

293. Ribak, J., et al. 1995. Common Accidents among Airport Ground Personnel. *Aviation Space Environmental Medicine* 66(12): 1188–1190.

294. Scheuch, K., et al. 2003. Evaluation Criteria for Aircraft Noise. *Review of Environmental Health* 18(3): 185–201.

295. Shouksmith, G. 2003. Human Factors in Air Traffic Control: Problems at the Interfaces. *Percep. Motor Skills* 97(2): 533–536.

296. Zeier, H. 1994. Workload and Psycho-physiological Stress Reactions in Air Traffic Controllers. *Ergonomics* 37(3): 525–539.

491: Postal Service

READINGS IN HISTORY

297. Adams, G. 1988. How Pioneers Got Their Mail. *Pacific Northwesterner* 33(2): 17–24.

298. Allen, J. B. E. 1990. Skiing Mailmen of Mountain America: U.S. Winter Postal Service in the 19th Century. *Journal of the West* 29(2): 76–86.

299. Atwood, R. A. 1986. Routes of Rural Discontent: Cultural Contradictions of Rural Free Delivery in Southeastern Iowa, 1899–1917. *Annals of Iowa* 48(5–6): 264–273.

300. Baarslag, K. 1945. *History of the National Federation of Post Office Clerks*. Washington, DC: National Federation of Post Office Clerks.

301. Baxter, V. K. 1984. Organizational Change, Technology & Labor in the Post Office. Dissertation, University of Wisconsin, Madison.

302. Carpenter, D. P. 2000. State Building through Reputation Building. *Studies in American Political Development* 14(2): 121–155.

303. Cullinan, G. 1973. *The United States Postal Service*. New York: Praeger.

304. Eigenheer, R. A. 1982. Eastward the Frontier: Historic Nevada Post Office Locations, 1860–1910. *Nevada Historical Society Quarterly* 25(4): 315–326.

305. Foley, M. S. 1997. A Mission Unfulfilled: The Post Office and the Distribution of Information in Rural New England, 1821–1835. *Journal of the Early Republic* 17(4): 611–650.

306. Frederick, J. V. 1937. The Holladay Overland Mail and Express Company. Dissertation, University of Oklahoma.

307. Germano, T. J. 1983. Labor Relations in the United States Postal Service: A Sociological Perspective. Dissertation, City University of New York.

308. Hochfelder, D. 2000. A Comparison of the Postal Telegraph Movement in Great Britain and the United States, 1866–1900. *Enterprise & Society* 1(4): 739–761.

309. John, R. R. 1989. Managing the Mails: The Postal System, Public Policy, and American Political Culture, 1823–1836. Dissertation, Harvard University.

310. ———. 1995. *Spreading the News: The American Postal System from Franklin to Morse.* Cambridge, MA: Harvard University Press.

311. Johnstone, D. B. 1992. *Postal History of Burlington, Vermont: The First 100 Years.* Burlington, VT: Queen City Publishers.

312. Kielbowicz, R. B. 1985. Speeding the News by Postal Express, 1825–1861. *Social Science Journal* 22(1): 49–63.

313. ——. 1986. Origins of the Second-class Mail Category and the Business of Policymaking, 1863–1879. *Journalism Monographs* 96: 1–26.

314. ——. 1989. *News in the Mail: The Press, Post Office, and Public Information, 1700–1860s*. Westport, CT: Greenwood Press.

315. ——. 1993. Origins of the Junk-mail Controversy: A Media Battle over Advertising and Postal Policy. *Journal of Public Policy History* 5(2): 248–272.

316. ——. 1994. Cost Accounting in the Service of Policy Reform: Postal Rate Making, 1875–1926. *Social Science Quarterly* 75(2): 284–299.

317. Kolva, H. J. 1987. Neither Sleet nor Snow . . . the U.S. Postal Service Comes to Colorado. *Colorado Heritage* 1: 2–6.

318. Lavender, A. D. 1999. Miami Beach's First Post Office. *South Florida History* 27(2): 10–15.

319. Lawson, L. and Kielbowicz, R. B. 1988. Library Materials in the Mail: A Policy History. *Library Quarterly* 58(1): 29–51.

320. Leutz, W. N. 1981. Aging and Technological Change: A Case Study of the U.S. Postal Service. Dissertation, Brandeis University.

321. Margolis, R. J. 1983. In Rural America the Post Office Remains a Mecca. *Smithsonian* 14(4): 62–68.

322. Matthews, J. T. 1999. Pushing the Mules through Guadalupe Pass: Pinery Station on the Butterfield Overland Mail. *Journal of Big Bend Studies* 11: 109–118.

323. Olds, K. B. 1995. The Challenge to the U.S. Postal Monopoly, 1839–1851. *Catao Journal* 15(1): 1–24.

324. Owens, P. A. 1989. The Overland Mail in Wyoming. *Annals of Wyoming* 61(2): 13–19.

325. Patterson, J. W. 1966. The Post Office in Early Minnesota. *Minnesota History* 40(2): 78–89.

326. Pearson, E. F. and Miles, W. 1980. Disinfection of Mail in the United States. *Bulletin of the History of Medicine* 54(1): 111–124.

327. Phillips, P. J. 1990. Never a Safe Road: Postal Communications in the Southwest Territory. *Journal of East Tennessee History* 62: 18–32.

328. Rachleff, P. 1982. Working the Fast Lane: Jobs, Technology and Scientific Management in the U.S. Postal Service. *Radical America* 16(1–2): 79–97.

329. Rich, W. E. 1924. *The History of the U.S. Post Office to the Year 1829*. Cambridge, MA: Harvard University Press.

330. Shotridge, J. R. 1974. The Post Office Frontier in Kansas. *Journal of the West* 13(3): 83–97.

331. Thomson, J. P. 1982. Spokane's Last Mail Dispatch by Stagecoach. *Pacific Northwesterner* 26(1): 11–16.

332. Vittes, M. E. 1984. Postal Service and the Public: A Case Study in Public Policy. Dissertation, University of Massachusetts.

333. ——. 1985. Value Tradeoffs and Productivity: The Transpoliticization of the U.S. Postal Service. *Policy Studies Review* 4(3): 504–511.

334. Werner, S. B. 1982. The Political Reversibility of Administrative Reform: A Case Study of the United States Postal Service. *Public Administration* 60(3): 320–338.

335. Wright, M. D. 1993. Privatization, the State, and Public Interest Trade Unionism: A Comparison of Canadian and American Postal Services. Dissertation, Stanford University.

336. Wriston, J. C. 1986. Vermont Post Office Locations. *Vermont History News* 37(6): 129–131.

READINGS IN HEALTH AND SAFETY

337. Baxter, V. 1996. Assaultive Violence in the U.S. Post Office. *Work & Occupations* 23 (3): 277–296.

338. Cahill, J. and Landsbergis, P. A. 1996. Job Strain among Post Office Mail Handlers. *International Journal of Health Service* 26(4): 731–750.

339. Cole, L. A. 2003. Persistence of a Mock Bio-agent in Cross-contaminated Mail and Mailboxes. *Journal of Public Health Management & Practice* 9(5): 357–360.

340. Fregert, S., et al. 1975. Chromate in Postage Stamps. *Contact Dermatitis* 1(5): 328–329.

341. Menne, T. 1983. Frictional Dermatitis in Post Office Workers. *Contact Dermatitis* 9(2): 172–173.

342. Sobti, A., et al. 1998. Occupational Physical Activity and Long-Term Risk of Muskoskeletal Symptoms: A National Survey of Post Office Pensioners. *American Journal of Industrial Medicine* 32(1): 76–83.

492: Couriers and Messengers

READINGS IN HISTORY

343. Anonymous. 1939. The Birth of the Express Business. *Business History Society Bulletin* 13: 59–63.

344. Culley, H. 2001. *The Immortal Class: Bike Messengers and the Cult of Human Power*. New York: Villard.

345. Davis, C. C. 1968. Effie Goldsborough: Confederate Courier. *Civil War Times Illustrated* 7(1): 29–31.

346. Downey, G. 2003. Telegraph Messenger Boys. *Professional Geographer* 55(2): 134–145.

347. Grossman, P. Z. 1992. Golden Silence: Why the Express Chose Not to Incorporate. *Business & Economic History* 21(2): 300–306.

348. John, R. R. 1986. Private Mail Delivery in the U.S. During the 19th Century. *Business & Economic History* 15: 135–147.

349. Langley, H. D. 1971. Early Diplomatic Couriers. *Foreign Service Journal* 48(10): 6–10.

350. Morlock, E. K., et al. 2000. *The Parcel Service Industry in the U.S.: Its Size and Role in Commerce*. Philadelphia: University of Pennsylvania, School of Engineering and Applied Science, Systems Engineering Department.

READINGS IN HEALTH AND SAFETY

351. Dennerlein, J. T. and Meeker, J. D. 2002. Occupational Injuries among Boston Bicycle Messengers. *American Journal of Industrial Medicine* 42(6): 519–525.

493: Warehousing and Storage

READINGS IN HISTORY

352. Alexander, F. M. 1989. Cabinet Warehousing in the Southern Atlantic Ports, 1783–1820. *Journal of Early Southern Decorative Arts* 15(2): 1–42.

353. Baker, J. D. 1972. Anonymity and American Architecture. *Historic Preservation* 24(3): 12–17.

354. Blunt, R. H. 1964. Lynchburg's Tobacco Warehouses. *Virginia Cavalcade* 14(3): 16–21.

355. Eaton, L. K. 1982. Warehouses and Warehouse Districts in Mid-American Cities. *Urban History Review* 11(1): 17–26.

356. Garvin, D-B. 1992. Warde's Stone Warehouse: Its Early History. *Historical New Hampshire* 47(3–4): 95–111.

357. Korpela, J. and Tuominen, M. 1996. A Decision Aid in Warehouse Site Selection. *International Journal of Production Economics* 45(1): 169–180.

358. Lee, G. A. 1937. The Historical Significance of the Chicago Grain Elevator System. *Agriculture History* 11: 16–32.

359. Rosenberger, H. L. 1994. Public Refrigerated Warehousing of Candy. *Manufacturing Confectioner* 74(6): 85.

360. Schafran, B. F. and Armstrong, C. W. 1991. Repair and Maintenance of Refrigerated Warehouse Floors. *Journal of Protective Coatings & Linings* 8(3): 60–64.

361. Schwartz, H. 1978. *The March Inland: Origins of the ILWU Warehouse Division, 1934–1938.* Los Angeles: University of California, Institute of Industrial Relations.

362. Stoeckle, R. 2001. Refrigerated Warehouse Operation under Real Time Pricing. Dissertation, University of Wisconsin, Madison.

363. Tschofen, C. 1997. The Finest Warehouse West of Chicago: Bishop's Block in Dubuque. *Iowa Heritage Illustrated* 78(2): 88–95.

364. Ulen, T. S. 1982. The Regulation of Grain Warehousing and Its Economic Effects. *Agricultural History* 56(1): 194–214.

READINGS IN HAZARDOUS WASTE MANAGEMENT AND POLLUTION PREVENTION

365. Dipoma, L. *Management of Bulk Petroleum Products, Storage, and Distribution Facilities* (DOD Directive 4140.25). Washington, DC: Assistant Secretary of Defense.

366. Dodds, G. E. 1990. Toxicity of Wastewaters: Bulk Petroleum Storage and Distribution Centers. Dissertation, University of Melbourne.

367. U.S. Department of the Interior. n.d. *Pollution Prevention Handbook: Shipping and Receiving Operations* (No. 24). Washington, DC: U.S. Department of the Interior.

READINGS IN HEALTH AND SAFETY

368. Cosgrave, J. 1997. Estimating the Capacity of Warehouses. *Disasters* 21(2): 155–165.

369. Garg, A. and Saxena, U. 1985. Physiological Stresses in Warehouse Operations with Special Reference to Lifting Techniques and Gender. *AIHA Journal* 46(2): 53–59.

370. Grant, G. and Drysdale, D. 1995. Numerical Modeling of Early Flame Spread in Warehouse Fires. *Fire Safety Journal* 24(3): 247–278.

371. Long, A. 2001. Storage and Warehouse Areas Can Be Hazardous Working Environments. *Premises & Facilities Management.*

372. Ljungberg, A. S., et al. 1989. Occupational Lifting by Nursing Aides and Warehouse Workers. *Ergonomics* 32(1): 59–78.

373. Miles, S. D. and Cox, G. 1996. Prediction of Fire Hazards Associated with Chemical Warehouses. *Fire Safety Journal* 27(4): 265–287.

374. Revsbech, P. and Andersen, G. 1987. Storage Mite Allergy among Grain Elevator Workers. *Allergy* 42(6): 423–429.

375. Sherman, S. J. 1992. Warehouse Worker's Headache: Carbon Monoxide Poisoning from Propane-fueled Forklifts. *AIHA Journal* 53(6): A298.

376. St. Vincent, M., et al. 2005. Work Factors Affecting Manual Materials Handling in a Warehouse Superstore. *International Journal of Industrial Ergonomics* 35(1): 33–46.

Sector 51: Information

511: Publishing Industries (Except Internet)
5111: Newspaper, Periodical, Book, and Directory Publishers
5112: Software Publishers

READINGS IN HISTORY

1. Ancilla, S. M. 1959. Catholic Polish Book Publishing in U.S. 1871–1900. *Polish American Studies* 16(1/2): 1–12.
2. Benton, M. M. 1998. The Elite Book in America: Fine Book Publishing, 1920–1932. Dissertation, University of California, Berkeley.
3. Blumenthal, J. 1977. *The Printed Book in America*. Boston: Godine.
4. Bullock, P. L. 1981. The Afro-American Periodical Press, 1838–1909. Dissertation, Louisiana State University Press.
5. Carroll, K. G. 1986. Sterling, Campbell and Albright: Textbook Publishers, 1861–1865. *North Carolina Historical Review* 63(2): 169–198.
6. Christianson, E. B. 1972. Mergers in the Publishing Industry, 1958–1970. *Journal of Library History* 7(1): 5–32.
7. Clark, C. E. 1990. The Newspapers of Provincial America. *Proceedings of the American Antiquarian Society* 100(2): 367–389.
8. Dickey, R. 1965. The Role of the Regional Book. *Journal of Arizona History* 6(3): 155–161.

9. Duino, R. A. 1982. The Cleveland Book Trade, 1819–1912: Leading Firms and Outstanding Men. Dissertation, University of Pittsburgh.

10. Dykema-VanderArk, A. M. 1999. This Most Democratic Province of the Republic of Letters: Autobiography and Periodical Publishing in Turn of the Century America. Dissertation, Michigan State University.

11. Eddy, J. 2003. Bookwomen: Creating an Empire in Children's Book Publishing, 1919–1939. Dissertation, University of Rochester.

12. Epstein, J. 2001. *Book Business: Publishing Past, Present and Future*. New York: Norton.

13. Goodrich, J. W. 2002. Duane Evans Lyon: A Sketch of an Artist. *Missouri Historical Review* 97(1): 59–72.

14. Greenberg, G. S. 1990. Isaac N. Whiting: Pioneer Bookseller and Publisher of Columbus, Ohio. *Publishing History* 28: 57–78.

15. Hanff, P. E. 1998. Way & Williams, Publishers, Chicago, 1895–1898. *Printing History* 20(2): 26–35.

16. Haugland, A. 2000. Book Propaganda: Edward L. Bernay's 1930 Campaign against Dollar Books. *Book History* 3: 231–252.

17. Helfand, W. H. 1978. Pharmaceutical and Medical Valentines. *Pharmacy in History* 20(3): 101–110.

18. Johnson, S. K. 1971. Sociology of Christmas Cards. *Trans-action* 8(3): 27–29.

19. Joyce, D. F. 1983. *Gatekeepers of Black Culture: Black-Owned Book Publishing in the United States, 1817–1981.* Westport, CT: Greenwood Press.

20. Kaur-Kasior, S. 1988. The Treatment of Culture in Greeting Cards: A Content Analysis. Dissertation, Bowling Green State University.

21. Lambert, L. F. 1981. The Seasonal Trade: Gift Cards and Chromolithography in America, 1874–1910. Dissertation, George Washington University.

22. Lazarsfeld, P. F. 1942. The Daily Newspaper and Its Competitors. *Annals of the American Academy of Political & Social Science* 219: 32–43.

23. Lehman, D. 1988. On the Publishing Front. *Partisan Review* 55(3): 371–386.

24. Lupfer, E. 2001. Before Nature Writing: Houghton, Mifflin and Company and the Invention of the Outdoor Book, 1800–1900. *Book History* 4: 177–204.

25. Neavill, G. B. 1977. Book Distribution and Marketing in the United States: A Review Article. *Library Quarterly* 47(1): 62–70.

26. Neiva, E. M. 1996. Chain Building: The Consolidation of the American Newspaper Industry, 1953–1980. *The Business History Review* 70(1): 1–42.

27. Noble, A. G. 2001. The Evolution of a Professional Journal: The First Thirty Years of Pioneer America/Material Culture. *Pioneer America Society Transactions* 24: 1–9.

28. O'Dowd, S. C. 2001. Frances Whipple and the Wampanoag: A 19th Century New England Factory Magazine. *Rhode Island History* 59(3): 67–83.

29. Petersen, W. J. 1967. Postcard Holiday Greetings. *Palimpsest* 48(12): 569–584.

30. Ray, R. H. 1951. Competition in the Newspaper Industry. *Journal of Marketing* 15(4): 444–456.

31. Remer, R. 1994. Preachers, Peddlers, and Publishers: Philadelphia's Backcountry Book Trade, 1800–1830. *Journal of the Early Republic* 14(4): 497–522.

32. Ristow, W. W. 1978. Worlds of Christmas Greetings. *Quarterly Journal of the Library of Congress* 35(4): 234–241.

33. Seeley, M. E. 1996. Seasons' Greetings from the White House: Presidential Words and Pictures. *Prologue* 28(4): 304–313.

34. Shank, B. 2004. *A Token of My Affection: Greeting Cards and American Business Culture*. New York: Columbia University Press.

35. Smith, A. 1980. *Goodbye Gutenberg: The Newspaper Revolution of the 1980s*. New York: Oxford University Press.

36. Sirianni, F. J. 1985. The Book Industry in Transition: A Study of Capital and Culture. Dissertation, State University of New York, Stony Brook.

37. Sullivan, P. W. 1974. G. D. Crain and the Founding of *Advertising Age*. *Journalism History* 1(3): 94–95.

38. Tebbel, J. 1972. *A History of Book Publishing in the United States. Vol. 1: The Creation of an Industry, 1630–1865*. New York: Bowker.

39. ———. 1975. *A History of Book Publishing in the United States. Vol. 2: The Expansion of an Industry, 1865–1919*. New York: Bowker.

40. ———. 1978. *A History of Book Publishing in the United States. Vol. 3: The Golden Age Between Two Wars, 1920–1940*. New York: Bowker.

41. ———. 1981. *A History of Book Publishing in the United States. Vol. 4: The Great Change, 1940–1980*. New York: Bowker.

42. ———. 1987. *Between Covers: The Rise and Transformation of American Book Publishing*. New York: Oxford University Press.

43. Todd, R. 1971. We Have Several Tons of Excess Sequins: The Hallmark Process. *Audience* 1(2): 4–11.

44. Tritter, T. 2005. The Press and Its Place: How Newspapers Fit into Urban History. *Journal of Urban History* 31(4): 571–578.

45. Trumbold, C. and Day, J. 1990. Mosby Year Book Inc. through 84 Years of Scientific Publishing. *Journal of Laboratory and Clinical Medicine* 116(5): 751–752.

46. West, J. L. W. 1990. Book Publishing 1835–1900. *Papers of the Bibliographical Society of America* 84(4): 357–375.

47. Winship, M. 1992. BAL and American Book Trade History. *Papers of the Bibliographical Society of America* 86(2): 147–155.

48. Woodress, J. L., ed. 1973. *Essays Mostly on Periodical Publishing in America: A Collection in Honor of Clarence Gohdes*. Durham, NC: Duke University Press.

49. Woods, M. 1989. The First American Architectural Journals: The Profession's Voice. *Journal of the Society of Architectural Historians* 48(2): 117–138.

READINGS IN HEALTH AND SAFETY

50. Bertazzi, P. A. and Zocchetti, C. 1980. A Mortality Study of Newspaper Printing Workers. *American Journal of Industrial Medicine* 1(1): 85–97.

51. Cook, C. and Page, E. 2001. Exposures of Workers at a Magazine Printing Company. *Applied Occupational & Environmental Hygiene* 16(6): 639–644.

52. Jensen, R. A. and Laxen, D. P. 1985. Sources of Lead in Urban Dust: Identification of a Contribution from Newspaper Printworks. *Science of the Total Environment* 46: 19–27.

53. Kornith, G., et al. 2005. Visible and Subclinical Skin Changes in Male and Female Dispatch Department Workers of Newspaper Printing Plants. *Skin Research Technology* 11(2): 132–139.

54. Liu, Y. H., et al. 2002. Increased Morbidity from Nasopharyngeal Carcinoma and Chronic Pharyngitis or Sinusitis among Workers at a Newspaper Printing Company. *Occupational & Environmental Medicine* 59(1): 18–22.

55. Luce, D., et al. 1997. Cancer Mortality among Magazine Printing Works. *Occupational & Environmental Medicine* 54(4): 264–267.

56. Purdham, J. T., et al. 1993. The Evaluation of Exposure of Printing Trade Employees to Polycyclic Aromatic Hydrocarbons. *Annals of Occupational Hygiene* 37(1): 35–44.

57. Yakes, B., et al. 1991. Occupational Skin Disease in Newspaper Pressroom Workers. *Journal of Occupational Medicine* 33(6): 711–717.

512: Motion Picture and Sound Recording Industries
5121: Motion Picture and Video Industries
5122: Sound Recording Industries

READINGS IN HISTORY

58. Banks, J. E. 1991. The Historical Development of the Video Music Industry: A Political Economic Analysis. Dissertation, University of Oregon.

59. Bell, S. 2003. From Ticket Booth to Screen Tower: An Architectural Study of Drive-in Theatres in the Baltimore–Washington, D.C.–Richmond Corridor. *Perspectives in Vernacular Architecture* 9: 215–227.

60. Butko, B. and Shiffer, R. 1994. Moonbeams and B-movies: The Rise and Fall of the Drive-in Theater. *Pennsylvania Heritage* 20(3): 16–23.

61. Carlson, W. B. and Gorman, M. E. 1990. Understanding Invention as a Cognitive Process: The Case of Thomas Edison and Early Motion Pictures, 1888–1891. *Social Studies of Science* 20(3): 387–430.

62. Cassady, R. 1933. Some Economic Aspects of Motion Picture Production and Marketing. *The Journal of Business of the University of Chicago* 6(2): 113–131.

63. Cohen, M. M. 1994. Forgotten Audiences in the Passion Pits: Drive-in Theatres and Changing Spectator Practices in Postwar America. *Film History* 6(4): 470–486.

64. Doerksen, C. J. 2002. Rough Music: The Cultural Politics of American Radio Broadcasting, 1920–1932. Dissertation, Princeton University.

65. Emmons, W. 1992. Public Policy and the Evolution of Cable Television, 1950–1990. *Business & Economic History* 21: 182–191.

66. Galambos, L. and Abrahamson, E. J. 2002. *Anytime, Anywhere: Entrepreneurship and the Creation of a Wireless World.* New York: Cambridge University Press.

67. Gronow, P. 1983. The Record Industry: The Growth of a Mass Medium. *Popular Music* 3: 53–75.

68. Harrod, F. S. 1993. Managing the Medium: The Navy and Motion Pictures before World War I. *Velvet Light Trap* 31: 48–58.

69. Herzog, C. 1981. The Movie Palace and the Theatrical Sources of Its Architectural Style. *Cinema Journal* 20(2): 15–37.

70. Horning, S. S. 2000. Chasing Sound: The Culture and Technology of Recording Studios in Postwar America. *American Icon* 6: 100–118.

71. Johnston, W. A. 1926. The Structure of the Motion Picture Industry. *Annals of the American Academy of Political & Social Science* 128: 20–29.

72. Killmeier, M. A. 2003. Mobile Media, Mobile Listeners: Automotive Radio and 1950s Radio Broadcasting. Dissertation, University of Iowa.

73. Kraft, J. P. 1991. Stage or Studio: American Musicians and Sound Technology in Motion Pictures, Radio, and Recording 1890–1946. *Business & Economic History* 20: 18–22.

74. ——. 1996. *Stage to Studio: Musicians and the Sound Revolution, 1890–1950.* Baltimore, MD: Johns Hopkins University Press.

75. Kuiper, J. B. 1968. The Historical Value of Motion Pictures. *American Archivist* 31(4): 385–390.

76. Laemmle, C. 1989. The Business of Motion Pictures. *Film History* 3(1): 47–71.

77. Lindsey, S. S. 1993. Screening Spaces: Women and Motion Pictures in America, 1908–1917. Dissertation, New York University.

78. Luther, R. 1950. Marketing Aspects of Drive-in Theatres. *Journal of Marketing* 15(1): 41–47.

79. Mannes, G. 1996. The Birth of Cable TV. *American Heritage of Invention & Technology* 12(2): 42–50.

80. Merchant, T. 2001. Drive-in Theatres of Northeast Wisconsin. *Voyageur* 18(1): 50–56.

81. Morton, D. 2000. *Off the Record: The Technology and Culture of Sound Recording in America*. New Brunswick, NJ: Rutgers University Press.

82. Morton, D. L. 1993. The Rusty Ribbon: John Herbert Orr and the Making of the Magnetic Recording Industry, 1945–1960. *The Business History Review* 67(4): 589–622.

83. Mullen, M. 2003. *The Rise of Cable Programming in the U.S.: Revolution or Evolution?* Austin: University of Texas Press.

84. Murray, J. B. 2001. *Wireless Nation: The Frenzied Launch of the Cellular Revolution in America.* Cambridge, MA: Perseus.

85. O'Connell, J. 1992. The Fine Tuning of a Golden Ear: High-end Audio and the Evolutionary Model of Technology. *Technology and Culture* 33(1): 1–37.

86. Odlum, F. B. 1947. Financial Organization of the Motion Picture Industry. *Annals of the American Academy of Political & Social Science* 254: 18–25.

87. Phelan, J. J. 2001. Motion Pictures as a Phase of Commercialized Amusement in Toledo, Ohio. *Film History* 13(3): 231–328.

88. Quinn, M. J. 1998. Early Feature Distribution and the Development of the Motion Picture Industry: Famous Players and Paramount, 1912–1921. Dissertation, University of Wisconsin, Madison.

89. Reddick, D. B. 1986. Movies under the Stars: A History of the Drive-in Theatre Industry, 1933–1983. Dissertation, Michigan State University.

90. Riley, N. 1985. Video Wars: A Phenomenon of the Music Video Industry. *Crisis* 92(7): 27–31.

91. Russo, A. T. 2004. Roots of Radio's Rebirth: Audiences, Aesthetics, Economics, and Technologies of American Broadcasting, 1926–1951. Dissertation, Brown University.

92. Schicke, C. A. 1974. *Revolution in Sound: A Biography of the Recording Industry*. Boston: Little, Brown.

93. Smith, L. F. 1979. *Perspectives on Radio and Television: An Introduction to Broadcasting in the U.S.* New York: Harper & Row.

94. Smith, L. F., et al. 1995. *Electronic Media and Government: The Regulation of Wireless and Wired Communication in the U.S.* White Plains, NY: Longman.

95. Staiger, J. 1984. Combination and Litigation: Structures of U.S. Film Distribution, 1896–1917. *Cinema Journal* 23(2): 41–72.

96. Sterne, J. 2003. *The Audible Past: Cultural Origins of Sound Reproduction.* Durham, NC: Duke University Press.

97. Tallmadge, W. H. 1979. Equipment Failure and Audio Distortion in the Acoustical Recording and Remastering of Early Jazz. *Journal of Jazz Studies* 5(2): 61–75.

98. Kodak. 2006. *Safe Handling, Storage and Destruction of Nitrate-based Motion Picture Films.* http://motion.kodak.com/motion/uploadedFiles/eknec_documents_72_0900688a80102572_H-182.pdf (accessed 29 October 2008).

READINGS IN HAZARDOUS WASTE MANAGEMENT AND POLLUTION PREVENTION

99. Keiler, J. A., et al. 1976. Persulfate/Quinine Bleach: Environmental and Economics Aspects. *SMPTE Journal* 95: 220–223.

100. Kodak. 2006. *Safe Handling, Storage and Destruction of Nitrate-based Motion Picture Films.* http://motion.kodak.com/motion/uploadedFiles/eknec_documents_72_0900688a80102572_H-182.pdf (accessed 29 October 2008).

101. Tisch, T. 1976. Cleaning Solutions for a Clean Environment: Developments in Motion Picture Film Cleaning Technology. *SMPTE Journal* 104: 528–533.

READINGS IN HEALTH AND SAFETY

102. Bonacich, E. 2005. Labor and the Television Industry. *Social Semiotics* 15(3): 335–342.

103. Hirschowitz, B.I. 1967. Use of Motion Pictures in Teaching and Diagnosis. *Annals of the New York Academy of Sciences* 142(2): 455–460.

104. Johnson, B. R. 1980. General Occurrence of Stressful Reactions to Commercial Motion Pictures and Elements in Films Subjectively Identified as Stressors. *Psychological Reports* 47(3) (Part 1): 775–786.

105. Lanis, L., et al. 2003. The Smoke Free Movies Initiative in the U.S. *Epidemiology Prevention* 27(4): 247–250.

106. Lee, I. and Tyrer, P. 1980. Responses of Chronic Agoraphobics to Subliminal and Supraliminal Phobic Motion Pictures. *Journal of Nervous and Mental Disorders* 168(1): 34–40.

107. Moline, J. M., et al. 2000. Health Effects Evaluation of Theatrical Smoke, Haze, and Pyrotechnics. Report prepared for Equity-League Pension & Health Trust Funds.

108. Redelmeier, D. 2003. Association between Mortality and Occupation among Movie Directors and Actors. *American Journal of Medicine* 115(5): 400–403.

109. Root, W. S., et al. 1953. Motion Pictures of Interest to the Physiologist. *Journal of Applied Physiology* 6(5): 315–316.

110. Teschke, K., et al. 2003. Exposure to Atmospheric Effects in the Entertainment Industry. *Journal of Occupational & Environmental Hygiene* 2(5): 277–284.

111. Varughese, S., et al. 2005. Effects of Theatrical Smokes and Fogs on Respiratory Health in the Entertainment Industry. *American Journal of Industrial Medicine* 47(5): 411–418.

513: Broadcasting and Telecommunications
5131: Radio and Television Broadcasting
5132: Cable and Other Subscription Programming
5133: Telecommunications

READINGS IN HISTORY

112. Agar, J. 2003. *Constant Touch: A Global History of the Mobile Phone*. Cambridge: Icon.

113. Albarran, A. B. and Pitts, G. G. 2001. *The Radio Broadcasting Industry*. Boston: Allyn & Bacon.

114. Baldwin, T. F. and McVoy, D. S. 1983. *Cable Communication*. Englewood Cliffs, NJ: Prentice-Hall.

115. Barnouw, E. 1968. *The Golden Web: A History of Broadcasting in the United States, 1933–1953*. New York: Oxford University Press.

116. Baughman, J. L. 1992. *The Republic of Mass Culture: Journalism, Filmmaking and Broadcasting in America since 1941*. Baltimore, MD: Johns Hopkins University Press.

117. Chisholm, B. F. 1988. The CBS Color Television Venture: A Study of Failed Innovation in the Broadcasting Industry. Dissertation, University of Wisconsin, Madison.

118. Coase, R. H. 1965. Evaluation of Public Policy Relating to Radio and Television Broadcasting. *Land Economics* 41(2): 161–167.

119. ——. 1979. Payola in Radio and Television Broadcasting. *Journal of Law & Economics* 22(2): 269–328.

120. Douglas, S. J. 1987. *Inventing American Broadcasting, 1899–1922*. Baltimore, MD: Johns Hopkins University Press.

121. Eisenmann, T. R. 2000. The U.S. Cable Television Industry, 1948–1995: Managerial Capitalism in Eclipse. *The Business History Review* 74(1): 1–40.

122. Engleman, R. 1996. *Public Radio and Television in America: A Political History*. Thousand Oaks, CA: Sage Publications.

123. Hausman, J. 1999. Cellular Telephone, New Products, and the CPI. *Journal of Business & Economic Statistics* 17(2): 188–194.

124. Hilmes, M. 1990. *Hollywood and Broadcasting: From Radio to Cable*. Urbana: University of Illinois Press.

125. Jome, H. L. 1925. Public Policy toward Radio Broadcasting. *The Journal of Land & Public Utility Economics* 1(2): 198–214.

126. Leonhirth, W. J. 1998. Selling Cable Television in the 1970s and 1980s: Social Dreams and Business Schemes. *American Journalism* 15(4): 95–124.

127. Lomax, K. S. 1999. Brought to You by Transcription: Portland's Radio History. *Oregon Historical Quarterly* 100(3): 320–327.

128. Maral, G. and Bousquet, M. 1986. *Satellite Communication Systems*. New York: Wiley.

129. Mitchell, B. M. and Vogelsang, I. 1991. *Telecommunications Pricing: Theory and Practice*. New York: Cambridge University Press.

130. Negrine, R. M., ed. 1985. *Cable Television and the Future of Broadcasting*. New York: St. Martin's Press.

131. O'Rourke, J. S. 1982. The Development of Color Television: A Study in the Free Market Process. *Journalism History* 9(3–4): 78–85.

132. Park, R. E. 1972. Cable Television, UHF Broadcasting, and FCC Regulatory Policy. *Journal of Law & Economics* 15(1): 207–231.

133. Reich, L. S. 1977. Research, Patents, and the Struggle to Control Radio: A Study of Big Business and the Uses of Industrial Research. *The Business History Review* 51(2): 208–235.

134. Roberts, P. 1998. The Quest for Public Television. *Annals of Wyoming* 70(3): 34–44.

135. Rye, J. 1988. On the Air: The Early Years of Radio Broadcasting in El Paso. *Password* 33(2): 65–72.

136. Schultz, D. L. 1985. A History of the Television Interview Program in Network Broadcasting in the United States. Dissertation, New York University.

137. Smethers, J. S. and Jolliffe, L. B. 2000. Singing and Selling Seeds: The Live Music Era on Rural Midwestern Radio Stations. *Journalism History* 26(2): 61–70.

138. Smulyan, S. 1994. *Selling Radio: The Commercialization of American Broadcasting, 1920–1934*. Washington, DC: Smithsonian Institution Press.

139. Sterling, C. H. and Kittross, J. M. 1978. *Stay Tuned: A History of American Broadcasting*. Mahwah, NJ: Erlbaum.

140. Smith, F. L. 1979. *Perspectives on Radio and Television: An Introduction to Broadcasting in the United States*. New York: Harper & Row.

141. Stockdell, R. P. 1983. The Evolution of the Country Radio Format. *Journal of Popular Culture* 16(4): 144–151.

142. VonSchilling, J. A. 1995. Television during World War II: Home Front Service, Military Success. *American Journalism* 12(3): 290–303.

READINGS IN HAZARDOUS WASTE MANAGEMENT AND POLLUTION PREVENTION

143. Horvath, A. 1999. Supply Chain Environmental Assessment of the Telecommunications Sector. In *Proceedings of the 1999 IEEE International Symposium on Electronics and the Environment, May 11–13, Danvers, MA*. IEEE.

144. Kasahara, M. and Koike, H. 1997. Environmental Protection and Related Technologies in Telecommunications Business. *NTT Review* 9(4): 24–29.

145. Snowden, K. G. 1994. Environmental Life Cycle Assessment of Face Plates Used in the Telecommunications Industry. In *Electronics and the Environment Proceedings*. IEEE.

READINGS IN HEALTH AND SAFETY

146. Ahlberg, J., et al. 2003. Self-reported Stress among Multiprofessional Media Personnel. *Occupational Medicine* 53(6): 403–405.

147. Aniolczyk, H. and Zymslony, M. 1994. Exposure to Electromagnetic Fields in Operators of Broadcast Radio Stations Emitting at 737 kHz to 1602 kHz. *International Journal of Occupational Medicine & Environmental Health* 7(1): 33–39.

148. Bhattacharya, S., et al. 2004. *Credit, Intermediation, and the Macroeconomy: Models and Perspective*. Oxford, UK: Oxford University Press.

149. Borer, A., et al. 2005. Cell Phones and Acinetobacter Transmission. *Emerging Infectious Disease* 11(7): 1160–1161.

150. Licitra, G., et al. 2004. Evaluation of the Electromagnetic Field Level Emitted by Medium Frequency AM Broadcast Stations. *Radiation Protection & Dosimetry* 111(4): 391–395.

151. Metal, A. and Ghahramani, B. 1994. The Injury Profile of a Large Telecommunication Company: A Statistical Summary. *Ergonomics* 37(10): 1591–1601.

152. Moulder, J. E., et al. 1999. Cell Phones and Cancer: What Is the Evidence for a Connection? *Radiation Research* 151(5): 513–531.

153. Namias, N., et al. 2000. Pathogenic Bacteria on Personal Pagers. *American Journal of Infection Control* 28(5): 387–388.

154. Nordenberg, T. 2000. Cell Phones and Brain Cancer. No Clear Connection. *FDA Consumer* 34(6): 19–23.

155. Oktay, M. F., et al. 2004. Occupational Safety: Effects of Workplace Radiofrequencies on Hearing Function. *Arch. Med. Res.* 35(6): 517–521.

156. Singh, D., et al. 2002. Bacterial Contamination of Hospital Pagers. *Infectious Control Hospital Epidemiol.* 23(5): 274–276.

514: Information Services and Data Processing Services

READINGS IN HISTORY

157. Bearman, D. 1991. *Archival Management of Electronic Records*. Pittsburgh: Archives & Museum Informatics.

158. Bennett, C. and Timbrell, G. T. 2000. Application Service Providers: Will They Succeed? *Information System Frontiers* 2(2): 195–212.

159. Bobinski, G. S. 1969. *Carnegie Libraries: Their History and Impact on American Public Library Development*. Chicago: American Library Association.

160. Bostick, S. L. 2001. The History and Development of Academic Library Consortia in the United States: An Overview. *Journal of Academic Librarianship* 27(2): 128–130.

161. Buckland, M. K. and Liu, Z. 1995. History of Information Science. *Annual Review of Information Science & Technology* 30: 385–416.

162. Cook, M. 1993. *Information Management and Archival Data*. Brookfield, VT: Gower.

163. Cook, T. 1997. What Is Past Is Prologue: A History of Archival Ideas Since 1898 and the Future Paradigm Shift. *Archivaria* 43: 17–63.

164. Davis, D. G., et al. 1989. *American Library History: A Comprehensive Guide to the Literature*. Santa Barbara: ABC-CLIO.

165. Dearstyne, B. W. 1993. *The Archival Enterprise: Modern Archival Principles, Practices and Management Techniques*. Chicago: American Library Association.

166. Dickson, G. W., et al. 1984. Key Information Systems Issues for the 1980s. *MIS Quarterly* 8(3): 135–159.

167. Hausman, J. A., et al. 2001. Cable Modems and DSL: Broadband Internet Access for Residential Customers. *The American Economic Review* 91(2): 302–407.

168. Huston, G. 1999. *ISP Survival Guide: Strategies for Running a Competitive ISP*. New York: Wiley.

169. Miles, W. D. 1982. *A History of the National Library of Medicine*. Bethesda, MD: NLM.

170. Niederman, F., et al. 1991. Information Systems Management Issues for the 1990s. *MIS Quarterly* 15(4): 475–500.

171. Rayward, W. B. 1996. The History and Historiography of Information Science: Some Reflections. *Information Processing & Management* 32(1): 3–17.

172. Rizenthaler, M. L. 1983. *Archives & Manuscripts Conservation: A Manual on Physical Care and Management*. Chicago: American Society of Archivists.

173. Schellenberg, T. R. 1965. *The Management of Archives*. New York: Columbia University Press.

174. Schruers, M. 2002. The History and Economics of ISP Liability for Third Party Content. *Virginia Law Review* 88(1): 205–264.

175. Van Slyck, A. A. 1995. *Free to All: Carnegie Libraries and American Culture, 1890–1920*. Chicago: University of Chicago Press.

176. Wiegan, W. A. 2000. American Library History Literature, 1947–1997: Theoretical Perspectives? *Libraries & Culture* 35(1): 4–34.

177. Wiegan, W. A. and Davis, D. G. 1994. *Encyclopedia of Library History*. New York: Garland Publishing.

Sector 52: Finance and Insurance

521: Monetary Authorities—Central Bank
522: Credit Intermediation and Related Activities
523: Securities, Commodity Contracts, and Other Financial Investments and Related Activities
524: Insurance Carriers and Related Activities
525: Funds, Trusts, and Other Financial Vehicles

READINGS IN HISTORY

1. Blackburn, R. 2003. *Banking on Death: or, Investing in Life: The History and Future of Pensions*. Brooklyn, NY: Verso.
2. Blinder, A. S. 2000. Central Bank Credibility: Why Do We Care? How Do We Build It? *The American Economic Review* 90(5): 1421–1431.
3. Bordo, M. C. and Jonung, L. 1999. *The Future of EMU: What Does the History of Monetary Unions Tell Us?* NBER Working Paper No. 7365. Cambridge, MA: National Bureau of Economic Research.
4. Brown, G. W. 2003. *Administration of Wills, Trusts and Estates*. London: Thomson Learning.
5. Calominis, C. W. 1990. Is Deposit Insurance Necessary: A Historical Perspective. *The Journal of Economic History* 50(2): 283–295.
6. Cobble, D. S. 1991. Organizing the Postindustrial Work Force: Lessons from the History of Waitress Unionism. *Industrial and Labor Relations Review* 44(3): 419–436.

7. Dunne, G. T. 1960. *Monetary Decisions of the Supreme Court.* New Brunswick, NJ: Rutgers University Press.

8. Friedman, M. and Schwartz, A. 1963. *Monetary History of the U.S., 1867–1960.* Princeton, NJ: Princeton University Press.

9. Galbraith, J. K. 1990. *A Short History of Financial Euphoria.* New York Penguin Books.

10. ——. 1995. *Money: Whence It Came, Where It Went.* Boston: Houghton Mifflin.

11. Goldsworth, J. 1995. Opinion: Is the History of Trusts Any Use Today? *Trusts and Trustees* 1(8): 2–3.

12. Gorton, G. 1985. Clearinghouses and the Origin of Central Banking in the United States. *The Journal of Economic History* 45(2): 277–283.

13. Hacker, J. S. 1998. The Historical Logic of National Health Insurance: Structure and Sequence in the Development of British, Canadian, and U.S. Medical Policy. *Studies in American Political Development* 12(1): 57–130.

14. Hixson, W. F. 1993. *Triumph of the Bankers: Money and Banking in the Eighteenth and Nineteenth Centuries.* London: Praeger.

15. Joskow, P. L. 1973. Cartels, Competition and Regulation in the Property-liability Insurance Industry. *The Bell Journal of Economics and Management Science* 4(2): 375–427.

16. Katz, S. N. 1985. Legal Change and Legal Autonomy: Charitable Trusts in New York, 1777–1893. *Law and History Review* 3(1): 51–89.

17. Kunreuther, H. and Roth, R. J. 1998. *Paying the Price: The Status and Role of Insurance against Natural Disasters in the U.S.* Washington, DC: Joseph Henry Press.

18. Langbein, J. H. 1995. The Contractarian Basics of the Law of Trusts. *Yale Law Journal* 105: 625–631.

19. Launie, J. J. 1969. The Supply Function of Urban Property Insurance. *The Journal of Risk and Insurance* 36(3): 269–283.

20. Maioni, A. 1998. *Parting at the Crossroads: The Emergence of Health Insurance in the U.S. and Canada.* Princeton, NJ: Princeton University Press.

21. Rosenberg, M. K. 1977. Historical Perspective of the Development of Rate Regulation of Title Insurance. *The Journal of Risk and Insurance* 44(2): 193–209.

22. Ross, H. L. 1980. *Settled Out of Court: The Social Process of Insurance Claims Adjustment.* Piscataway, NJ: Aldine Transaction.

23. Weare, W. B. 1993. *Black Business in the New South: A Social History of the North Carolina Mutual Life Insurance Company*. Durham, NC: Duke University Press.

24. Wilson, J. W., et al. 2003. *A History of Public Sector Pensions in the United States*. Philadelphia: University of Pennsylvania Press.

25. Zeiler, V. A. 1978. Human Values and the Market: The Case of Life Insurance and Death in 19th Century America. *The American Journal of Sociology* 84(3): 591–610.

26. ——. 1983. *Morals and Markets: The Development of Life Insurance in the United States*. Edison, NJ: Transaction.

Sector 53: Real Estate and Rental and Leasing

531: Real Estate
532: Rental and Leasing Services
533: Lessors of Nonfinancial Intangible Assets (Except Copyrighted Services)

READINGS IN HISTORY

1. Abione, F. A. 1950. Marketing Equipment by Leasing. *Journal of Marketing* 15(2): 205–211.
2. Brown, W. H. 1972. Access to Housing: The Role of the Real Estate Industry. *Economic Geography* 48(1): 66–78.
3. Clure, R. M. 1990. The Assessment of Theatrical Costume Management Systems for Non-profit Theatres and Costume Rental Agencies. Dissertation.
4. Dana, J. D. and Spier, K. E. 2001. Revenue Sharing and Vertical Control in the Video Rental Industry. *Journal of Industrial Economics* 49(3): 223–245.
5. Davis, C. E. and Furst, E. 1988. Cost Effective Quality Assurance of Rented Medical Equipment. *Journal of Clinical Engineering* 13(6): 421–425.
6. Devavrat, P. and Staelin, R. 1994. Rentals, Sales, and Buybacks: Managing Secondary Distribution Channels. *Journal of Marketing Research* 31 (3): 325–338.
7. Diewert, W. E. 1974. Intertemporal Consumer Theory and the Demand for Durables. *Econometrica* 42(3): 497–516.
8. Duvall, S. G. and Ash, R. L. 1986. Caveat Lessor: Leasing to Tax Exempt Entities. *The Journal of Real Estate Taxation* 14(1): 54–80.

9. Ely, R. T. 1932. Real Estate in the Business Cycle. *The American Economic Review* 22(1): 137–143.

10. Griffin, J. 2003. Underground Construction Equipment Gains Large Share of Rental Business. *Underground Construction* 58(4): 56–57.

11. Hastak, M. 2004. Regulation of the Rent-to-Own Industry: Implications of the Wisconsin Settlement with Rent-a-Center. *Journal of Public Policy & Marketing* 23(1): 96–98.

12. Helper, R. 1969. *Racial Policies and Practices of Real Estate Brokers*. Minneapolis: University of Minnesota Press.

13. Kyle, R. C., et al. 1999. *Property Management*. Dearborn, MI: Dearborn Real Estate Education.

14. Lacko, J. M., et al. 2002. Customer Experience with Rent-to-Own Transactions. *Journal of Public Policy & Marketing* 21(1): 126–138.

15. McGugan, V. J. and Caves, R. E. 1974. Integration and Competition in the Equipment Leasing Industry. *The Journal of Business* 47(3): 382–396.

16. Mooradian, R. M. and Yang, S. X. 2002. Commercial Real Estate Leasing, Asymmetric Information, and Monopolistic Competition. *Real Estate Economics* 30(2): 293–315.

17. Pachon, J., et al. 2006. Vehicle Fleet Planning in the Car Rental Industry. *Journal of Revenue and Pricing Management* 5(3): 221–236.

18. Palm, R. 1976. The Role of Real Estate Agents as Information Mediators in Two American Cities. *Geografiska Annaler. Series B, Human Geography* 58(1): 28–41.

19. Taff, C. A. 1950. Developments in the Leasing of Motor Truck Equipment. *Southern Economic Journal* 17(1): 50–54.

20. Varian, H. R. 2000. Buying, Sharing and Renting Information Goods. *The Journal of Industrial Economics* 48(4): 473–488.

21. Weiss, M. A. 1984. Researching the History of Real Estate Development. *Journal of Architectural Education* 41(3): 38–40.

22. ——. 1989. Real Estate History: An Overview and Research Agenda. *The Business History Review* 63(2): 241–282.

23. Zumpano, L. V. and Hooks, D. L. 1988. The Real Estate Brokerage Market: A Critical Reevaluation. *Real Estate Economics* 16(1): 1–16.

Sector 54: Professional, Scientific, and Technical Services

READINGS IN HISTORY

1. Allen, D. G. and McDermott, K. 1993. *Accounting for Success: A History of Price Waterhouse in America, 1890–1990*. Harvard Business School Press.
2. American Society of Landscape Architects. 1989. *American Landscape Architecture: Designers and Place*. Washington, DC: Preservation Press.
3. Barrett, M. 1926. *History and Development of Bookkeeping and Accounting in the Secondary Schools*. Boston: Boston University Press.
4. Bevan, B. W. 2000. An Early Geophysical Survey at Williamsburg. *Archaeological Prospection* 7(1): 51–58.
5. Bierer, B. W. 1955. *A Short History of Veterinary Medicine in America*. East Lansing, MI: Michigan State University Press.
6. Biggart, N. W. 1989. *Charismatic Capitalism: Direct Selling Organizations in America*. Chicago: University of Chicago Press.
7. Brown, R. A. 1968. *History of Accounting and Accountants*. Oxford: Routledge.
8. Buhler, D. A. 2006. Cadastral Survey Activities in the United States. *Surveying and Land Information Science* 66(2): 115–118.
9. Buhler, D. A. and Racette, J. G. 2002. BLM Cadastral Survey: The History, the Program, and the Partnerships. *Surveying and Land Information Science* 62(1): 15–20.
10. Calhoun, D. 1960. *The American Civil Engineer: Origins and Conflict*. Cambridge, MA: Harvard University Press.

11. Chua, W. F. 1986. Radical Developments in Accounting Thought. *The Accounting Review* 61(4): 601–632.

12. Churchill, C. W. 1953. *A History of Bookkeeping Instruction in the U.S. from 1635 to 1953*. Chico, CA: Chico State College.

13. Cohen, A. 1999. Turnover among Professionals: A Longitudinal Study of American Lawyers. *Human Resource Management* 38(1): 61–76.

14. Czech, K. P. 1996. *Snapshot: America Discovers the Camera*. Minneapolis, MN: Lerner.

15. Dahlbom, B. and Mathiassen, L. 1983. *Computers in Context: The Philosophy and Practice of Systems Design*. Cambridge, MA: Blackwell Publishing.

16. Dubin, J. A., et al. 1992. The Demand for Tax Return Preparation Services. *The Review of Economics & Statistics* 74(1): 75–82.

17. Dunlop, R. H. and Williams, D. J. 1996. *Veterinary Medicine: An Illustrated History*. New York: Mosby.

18. Fehrman, C. and Fehrman, K. 1987. *Postwar Interior Design, 1945–1960*. New York: Van Nostrand Reinhold.

19. Fisher, L. K. 1998. American Society of Notaries: History of a Legacy. *John Marshall Law Review* 31(3): 1001–1006.

20. Flesher, D. L., et al. 1996. The First Century of CPA. *Journal of Accountancy* 182.

21. Fogle, H. 1943. The History of CPA Legislation in the United States. Dissertation, New York: Niagara University.

22. Fox, S. R. 1997. *The Mirror Makers: A History of American Advertising and Its Creators*. Chicago: University of Illinois Press.

23. Friedman, A. L. and Cornford, D. S. 1989. *Computer Systems Development: History, Organization and Implementation*. New York: John Wiley & Sons.

24. Frishberg, N. 1975. Arbitrariness and Iconicity: Historical Change in American Sign Language. *Language* 51(3): 696–719.

25. Frosh, P. 2003. *The Image Factory: Consumer Culture, Photography and the Visual Content Industry*. Oxford, UK: Berg Publishers.

26. Harvey, P. D. A. 1980. *The History of Topographical Maps: Symbols, Pictures, and Surveys*. London: Thames & Hudson.

27. Heinz, J. P., et al. 1998. The Changing Character of Lawyer's Work: Chicago in 1975 and 1995. *Law & Society Review* 32(4): 751–776.

28. Hickman, J. C. 1998. Actuaries in History: The Wartime Birth of Operations Research. *North American Actuarial Journal* 2(4): 1–10.

29. Hollis, R. 1999. *Graphic Design: A Concise History*. London: Thames & Hudson.

30. Hower, R. M. 1978. *The History of an Advertising Agency*. New York: Arno Press.

31. Johnson, H. T. 1975. The Role of Accounting History in the Study of Modern Business Enterprise. *The Accounting Review* 50(3): 444–450.

32. Johnson, H. T. and Kaplan, R. S. 1991. *Relevance Lost: The Rise and Fall of Management Accounting*. Boston: Harvard Business School Press.

33. Kaplan, R. S. 1984. The Evolution of Management Accounting. *The Accounting Review* 59(3): 390–418.

34. Kats, P. 1929. *Early History of Bookkeeping by Double Entry*. New York: American Institute of Certified Public Accountants.

35. Kitch, E. W. 1977. The Nature and Function of the Patent System. *Journal of Law & Economics* 20(2): 265–290.

36. Klepper, S., et al. 1991. Expert Intermediaries and Legal Compliance: The Case of Tax Preparers. *Journal of Law & Economics* 34(1): 205–229.

37. Kyle, J. G. 1988. *Sign Language: The Study of Deaf People and Their Language*. Cambridge, UK: Cambridge University Press.

38. Layton, E. T. 1976. American Ideologies of Science and Engineering. *Technology and Culture* 17(4): 688–701.

39. Lears, T. J. J. 1994. *Fables of Abundance: A Cultural History of Advertising in America.* New York: Basic Books.

40. Long, J. E. and Caudill, S. B. 1987. The Usage and Benefits of Paid Tax Return Preparation. *National Tax Journal* 40: 35–46.

41. Lucie-Smith, E. 1983. *A History of Industrial Design*. London: Phaidon.

42. Marchand, R. 2001. *Creating the Corporate Soul: The Rise of Public Relations and Corporate Imagery in America.* Berkeley: University of California Press.

43. Massey, A. 1990. *Interior Design of the 20th Century*. London: Thames & Hudson

44. McKenna, C. D. 1995. The Origins of Modern Management Consulting. *Business & Economic History* 24(1): 51–58.

45. Meggs, P. B. 2006. *The History of Graphic Design*. New York: John Wiley & Sons.

46. Miceli, T. J., et al. 2002. Title Systems and Land Values. *The Journal of Law & Economics* 45: 565–582.

47. Mohr, J. C. 2000. American Medical Malpractice Litigation in Historical Perspective. *JAMA* 283(13): 1731–1737.

48. Monmonier, M. 1997. *Cartographies of Danger: Mapping Hazards in America*. Chicago: University of Chicago Press.

49. Moorhead, E. J. 1989. *The History of the Actuarial Profession in North America, 1809–1979*. Schamburg, IL: The Society of Actuaries.

50. Munday, J. 2001. *Introducing Translation Studies: Theories and Applications*. Oxford: Routledge.

51. Orvell, M. 2003. *American Photography*. New York: Oxford University Press.

52. Parker, R. H. 1965. Accounting History: A Select Bibliography. *Abacus* 1(1): 62.

53. ——. 1993. The Scope of Accounting History: A Note. *Abacus* 29(1): 106.

54. Pickles, J. A. 2004. *History of Spaces: Cartographic Reason, Mapping and the Geo-coded World*. New York, Oxford: Routledge.

55. Previts, G. J., et al. 1990. An Accounting Historiography: Subject Matter and Methodology. *Abacus* 26(2): 136.

56. Rabin, R. L. 1992. A Sociolegal History of the Tobacco Tort Litigation. *Stanford Law Review* 44(4): 853–878.

57. Rogers, J. R., et al. 1996. *Civil Engineering History: Engineers Make History*. New York: ASCE.

58. Romano, R. 1993. *The Genius of American Corporate Law*. Washington, DC: American Enterprise Institute.

59. Rosenberg, M. K. 1977. Historical Perspective of the Development of Rate Regulation of Title Insurance. *The Journal of Risk & Insurance* 44(2): 193–209.

60. Roth, L. M. 1982. *A Concise History of American Architecture*. Boulder, CO: Westview Press.

61. Rubin, P. H. and Bailey, M. J. 1994. The Role of Lawyers in Changing the Law. *The Journal of Legal Studies* 23(2): 807–831.

62. Seth, J. E. 1999. Notaries in the American Colonies. *John Marshall Law Review* 32(4): 863–886.

63. Skeel, D. A. 2001. *Debt's Dominion: A History of Bankruptcy Law in America*. Princeton, NJ: Princeton University Press.

64. Smith, C. R. 1987. *Interior Design in 20th Century America: A History*. New York: Harper & Row.

65. Stafford, H. M. 1959. History of Snow Surveying in the West. Paper presented at the Western Snow Conference, April 21, Reno, NV.

66. Steele, J. H. 1991. History of Veterinary Public Health in the U.S. *Reviews in Science & Technology* 10(4): 951–983.

67. Treece, M. and Binford, P. 1969. Bookkeeping and Accounting: Rules and Reasons in the History of Bookkeeping Instruction. *Business Education Forum* 24(1): 21–22.

68. Valera, M. G. 1998. The National Notary Association: A Historical Profile. *John Marshall Law Review* 31: 971–999.

69. Venuti, L. 1995. *The Translator's Invisibility: A History of Translation*. Oxford: Routledge.

70. Warren, C. 1990. *History of the American Bar*. Buffalo: William S. Hein Publishing.

71. Weaver, B. M. 1987. The History of Veterinary Anesthesia. *Veterinary History* 5(2): 43–57.

72. Williams, H. E. 1978. Paralegal: History, Current Status, Emerging Trends, and Guidelines for Implementing a Program. Dissertation, Morehead State University.

73. Yamey, B. S. 1949. Scientific Bookkeeping and the Rise of Capitalism. *The Economic History Review* 1(2-3): 99–113.

74. Yetmar, S. A. and Eastman, K. K. 2000. Tax Practitioner's Ethical Sensitivity: A Model and Empirical Examination. *Journal of Business Ethics* 26(4): 1573–1697.

75. Zilkoski, D. B., et al. 1992. Results of the General Adjustment of the North American Vertical Datum of 1988. *Surveying and Land Information Systems* 52(3): 133–149.

Sector 55: Management of Companies and Enterprises

READINGS IN HISTORY

1. Aharony, J. and Swary, I. 1981. Effects of the 1970 Bank Holding Company Act: Evidence from Capital Markets. *The Journal of Finance* 36(4): 841–853.
2. Alexander, D. L. 1985. An Empirical Test of the Mutual Forbearance Hypothesis: The Case of Bank Holding Companies. *Southern Economic Journal* 52(1): 122–140.
3. Black, F., et al. 1978. An Approach to the Regulation of Bank Holding Companies. *The Journal of Business* 51(3): 379–412.
4. Carow, K. A. and Larsen, G. A. 1997. The Effect of FDICIA Regulation on Bank Holding Companies. *Journal of Financial Research* 20(2): 159–174.
5. Davenport, T. H. 1998. Putting the Enterprise into the Enterprise System. *Harvard Business Review* 76(4): 121–133.
6. Demsetz, R. S. and Strahan, P. E. 1997. Diversification, Size and Risk at Bank Holding Companies. *Journal of Money, Credit & Banking* 29(3): 300–313.
7. Lawrence, R. J. 1967. *The Performance of Bank Holding Companies*. Washington, DC: Federal Reserve System.
8. Rice, A. K. 1963. *The Enterprise and Its Environment: A System Theory of Management Organization*. Oxford: Routledge.
9. Rosenzweig, P. M. and Singh, J. V. 1991. Organizational Environments and the Multinational Enterprise. *The Academy of Management Review* 16(2): 340–361.

Sector 56: Administrative and Support and Waste Management and Remediation Services

561: Administrative and Support Services

READINGS IN HISTORY

1. Abbott, G. 1908. The Chicago Employment Agency and the Immigrant Worker. *The American Journal of Sociology* 14(3): 289–305.

2. Avramidis, A. N., et al. 2005. Modeling Daily Arrivals to a Telephone Call Center. *Management Science* 50(7): 896–908.

3. Berger, C. J. 1990. Timesharing in the United States. *The American Journal of Comparative Law* 38 (Supp.): 131–151.

4. Braun, B. M. 1992. The Economic Contribution of Conventions: The Case of Orlando, Florida. *Journal of Travel Research* 30(3): 32–37.

5. Briceland, A. V. 1972. Ephraim Kirby: Pioneer of American Law Reporting, 1789. *The American Journal of Legal History* 16(4): 297–319.

6. Capon, N. 1982. Credit Scoring Systems: A Critical Analysis. *Journal of Marketing* 46(2): 82–91.

7. Conrath, D. W. and Hamilton, W. F. 1971. The Economics of Manpower Pooling. *Management Science* 18(2): B19–B29.

8. Cowan, R. and Gunby, P. 1996. Sprayed to Death: Path Dependence, Lock-in and Pest Control Strategies. *The Economic Journal* 106(436): 521–542.

9. Custis, V. 1912. Lumber Grading in the Pacific Northwest. *The Quarterly Journal of Economics* 26(3): 538–544.

10. Dekimpe, M. G., et al. 1997. Generalizing about Trade Show Effectiveness: A Cross-national Comparison. *Journal of Marketing* 61(4): 55–64.

11. Dunne, G. T. 1976. Proprietors—Sometimes Predators: Early Court Reporters. In *Supreme Court Historical Society Yearbook,* 61–72.

12. Edler, F. C. 1954. *The Armored Car Service: Its History and Importance in Our Banking and Industrial Structure*. Chicago: Federal Reserve Bank of Chicago.

13. Escalante, D. 1995. A Study of Employee Leasing as an Alternative to Traditional Hiring Methods. Dissertation, Mercy College, Dobbs Ferry, NY.

14. Estevao, M. and Lach, S. 1999. *The Evolution of the Demand for Temporary Help Supply Employment in the United States.* NBER Working Paper Series 7427. Washington, DC: National Bureau of Economic Research.

15. Garelis, A. 1996. Quality in the Call Center Industry—What We Can Learn from History. *Telemarketing* 14(8): 68–73.

16. Gartrell, R. B. 1994. *Destination Marketing for Convention and Visitor Bureaus*. Dubuque, IA: Kendall/Hunt Publishing.

17. Hoyle, L. H., et al. 1989. *Managing Conventions and Group Business*. East Lansing, MI: Educational Institute of the American Hotel & Motel Association.

18. Hunt, R. M. 2005. *A Century of Consumer Credit Reporting in America.* Philadelphia: Federal Reserve Bank of Philadelphia.

19. Ingells, J. E. and Reiley, H. E. 1978. *Landscaping: Principles and Practices*. New York: Van Nostrand Reinhold.

20. Kagan, R. A. 1984. The Routinization of Debt Collection. *Law & Society Review* 18(3): 323–372.

21. Kerin, R. A. and Cron, W. L. 1987. Assessing Trade Show Functions and Performance: An Exploratory Study. *Journal of Marketing* 51(3): 87–94.

22. Koole, G. and Mandelbaum, A. 2002. Queuing Models of Call Centers: An Introduction. *Annals of Operation Research* 113(1/4): 41–59.

23. Krumbein, W. C. 1924. Collection Agencies. *The University Journal of Business* 3: 48–67.

24. Laband, D. N. and Maloney, M. T. 1994. A Theory of Credit Bureaus. *Public Choice* 80(3/4): 275.

25. Lee, R. H. 1924. Serving the Motorist: The Work of the National Motorists' Association. *Annals of the American Academy of Political & Social Science* 116: 266–268.

26. Lehigh University. 1991. *Mailing Services: Policies and Procedures*. Bethlehem, PA: Lehigh University.

27. Leiderman, R. 1990. *The Telephone Book: How to Find, Get, Keep, and Develop Customers*. New York: McGraw-Hill.

28. Levery, M. 1998. Outsourcing Maintenance—A Question of Strategy. *Engineering Management Journal* 8(1): 34–40.

29. Leviton, C. 1938. Automobile Club Activities: The Problem from the Standpoint of the Bar. *Law and Contemporary Problems* 5(1): 11–21.

30. McCrie, R. D. 1988. The Development of the U.S. Security Industry. *Annals of the American Academy of Political & Social Science* 498(1): 23–33.

31. Michaud, G. and Paul, J. 1982. The Importance of Bodyguards. *MERIP Reports* 110: 29–31.

32. Moore, M. A. 1965. The Temporary Help Service Industry: Historical Development, Operation and Scope. *Industrial & Labor Relations Review* 18(4): 554–569.

33. Murthy, B. and Murrmann, S. K. 1993. Employee Leasing: An Alternative Staffing Strategy. *Cornell Hotel & Restaurant Administration Quarterly* 34(3): 18.

34. Pennella, I. 1993. The Evolution of the Court Reporting Profession. Dissertation, California State University, Long Beach.

35. Robbe, D. 2000. *Expositions and Trade Shows*. New York: Wiley.

36. Rogers, T. 2003. *Conferences and Conventions: A Global Business*. Boston: Butterworth-Heinemann.

37. Scott, F. T. 1931. The Stenographer in a Machine Age. *Peabody Journal of Education* 9(2): 98–101.

38. Scott, T. M. and McPherson, M. 1971. The Development of the Private Sector of the Criminal Justice System. *Law & Society Review* 6(2): 267–288.

39. Segal, L. W. and Sullivan, D. G. 1997. The Growth of Temporary Services Work. *The Journal of Economic Perspectives* 11(2): 117–136.

40. Shearing, C. D. and Stenning, P. C. 1981. Modern Private Security: Its Growth and Implications. *Crime and Justice* 3: 193–245.

41. Sheldon, P. J. 2000. The Tour Operator Industry: An Analysis. *International Library of Critical Writings in Economics* 121(1): 330–346.

42. Shuchman, P. 1969. Profit on Default: An Archival Study of Automobile Repossession and Resale. *Stanford Law Review* 22(1): 20–66.

43. Siu, G. K. W., et al. 2001. Assessing the Service Quality of Building Maintenance Providers: Mechanical and Engineering Services. *Construction Management & Economics* 19(7): 719–726.

44. Smith, E. N. 1924. Services of the American Automobile Association. *Annals of the American Academy of Political & Social Science* 116: 269–274.

45. South, N. 1988. *Policing for Profit: The Private Security Sector.* Newbury Park, CA: Sage Publications.

46. Stees, J. D. 1998. *Outsourcing Security: A Guide for Contracting Services.* Boston: Butterworth-Heinemann.

47. Taylor, C. R. 1980. The Nature of Benefits and Costs of Use of Pest Control Methods. *American Journal of Agricultural Economics* 62(5): 1007–1011.

48. Theodore, N. and Peck, J. 2002. The Temporary Staffing Industry: Growth Imperatives and Limits to Contingency. *Economic Geography* 78(4): 463–494.

49. Tombs, M. J. 1996. *The Bodyguard Training Manual.* Durham, UK: Professional Bodyguard Association.

50. Tsang, A. H. C. 2002. Strategic Dimensions of Maintenance Management. *Journal of Quality in Maintenance Engineering* 8(1): 7–39.

51. Tschetter, J. 1987. Producer Services Industries: Why Are They Growing So Rapidly? *Monthly Labor Review* 110(12): 31.

52. Weiss, R. P. 1986. Private Detective Agencies and Labor Discipline in the United States, 1855–1946. *The Historical Journal* 29(1): 87–107.

53. Wilde, E. 1994. Dust, Sweep and Clean—A Case History. *Work Study* 43(5): 5.

54. Yudof, M. G. 1974. Reflections on Private Repossession, Public Policy and the Constitution. *University of Pennsylvania Law Review* 122(4): 954–981.

READINGS IN WASTE MANAGEMENT

55. Barron, T. and Sutherland, L. 1999. Environmentally Preferable Janitorial Products: Issues and Opportunities. *P2 Pollution Prevention Review* 9(4): 17–26.

READINGS IN HEALTH AND SAFETY

56. Brown, N. J. 1990. *Health Hazards Manual for Custodians, Janitors, and Housekeepers.* Buffalo, NY: Cornell University Chemical Hazard Information Program.

57. Kondrashov, V., et al. 2005. Assessment of Lead Exposure Risk in Locksmiths. *International Journal of Environmental Research & Public Health* 2 (1) : pp 164–69.

562: Waste Management and Remediation Services

READINGS IN HISTORY

58. Amey, R. G. 1998. Low-Level Radioactive Waste Management, Disposal and Facility Siting Issues. Dissertation, University of Florida.

59. Antunes, G. E. and Halter, G. 1976. The Politics of Resource Recovery, Energy Conservation and Solid Waste Management. *Administration & Society* 8(1): 55–78.

60. Armbrister, M. 2001. Solid Waste Disposal in Southwest Virginia: A Regional Solution. *International Social Science Review* 76(1–2): 3–11.

61. Barth, E. F. 1990. *Stabilization and Solidification of Hazardous Wastes*. Park Ridge, NJ: Noyes Data Corp.

62. Brunette, D. A. 1992. Charles Valentine Riley and the Roots of Modern Insect Control. *Missouri Historical Review* 86(3): 229–247.

63. Colten, C. E. 1986. Industrial Wastes in Southeast Chicago: Production and Disposal, 1870–1970. *Environmental Review* 10(2): 93–105.

64. ——. 1988a. Historical Questions in Hazardous Waste Management. *Public Historian* 10(1): 7–20.

65. ——. 1988b. Industrial Middens in Illinois: The Search for Historical Hazardous Wastes, 1870–1980. *Industrial Archeology* 14(2): 51–61.

66. ——. 1994a. Chicago's Waste Lands: Refuse Disposal and Urban Growth, 1840–1990. *Journal of Historical Geography* 20(2): 124–142.

67. ——. 1994b. Creating a Toxic Landscape: Chemical Waste Disposal Policy and Practice, 1900–1960. *Environmental History Review* 18(1): 85–116.

68. Conn, W. D. 1978. Planning for Resource Recovery: Lessons from the California Experience. *Journal of the American Institute of Planners* 44(2): 200–208.

69. Corey, S. H. 1994. King Garbage: A History of Solid Waste Management in New York, 1881–1970. Dissertation, New York University.

70. Dahlsten, D. L. and Garcia, R., eds. 1990. *Eradication of Exotic Pests: Analysis with Case Histories*. New Haven, CT: Yale University Press.

71. Everett, J. W. and Applegate, D. 1995. Solid Waste Transfer Station Design. *Journal of Environmental Engineering* 121(1): 96–106.

72. Hall, A. 1993. The Development of Solid Waste Transfer Stations. *Wastes Management* 5: 1–15.

73. Ham, R. K. 1993. Overview and Implications of U.S. Sanitary Landfill Practice. *Air & Waste* 43(2): 187–190.

74. Hegdahl, T. A. 1973. *Solid Waste Transfer Stations: A State-of-the-Art Report on Systems Incorporating Highway Transportation*. Washington, DC. U.S. Environmental Protection Agency.

75. Howley, J. 1990. Justice for Janitors: The Challenge of Organizing in Contract Services. *Labor Research Review* 9(1): 61–72.

76. Kauffman, H. J. 1992. Locksmiths and Locks in Lancaster County. *Journal of the Lancaster County Historical Society* 94(2): 52–57.

77. Khan, A. M. 1987. Solid Waste Disposal with Intermediate Transfer Stations. *The Journal of the Operational Research Society* 38(1): 31–37.

78. Klatzky, S. R. 1970. Organizational Inequality: The Case of the Public Employment Agencies. *American Journal of Sociology* 76(3): 474–491.

79. Leavitt, J. W. 1980. The Wasteland: Garbage and Sanitary Reform in the 19th Century American City. *Journal of the History of Medicine & Allied Sciences* 35(4): 431–452.

80. Lund, H. F. 1993. *The McGraw-Hill Recycling Handbook*. New York: McGraw-Hill.

81. Lund, J. R., et al. 1994. Linear Programming for Analysis of Material Recovery Facilities. *Journal of Environmental Engineering* 120(5): 1082.

82. Luthy, R., et al. 1994. Premeditating Tar-contaminated Soils at Manufactured Gas Plant Sites. *Environmental Science & Technology* 28(6): 266A.

83. McGowan, W. P. American Wasteland: A History of America's Garbage Industry, 1880–1989. *Business & Economic History* 24(1): 155–163.

84. Melosi, M. V. 1981. *Garbage in the Cities: Refuse, Reform and the Environment, 1880–1980*. College Station: Texas A&M University Press.

85. ——. 2002. The Fresno Sanitary Landfill in an American Cultural Context. *Public Historian* 24(3): 17–35.

86. Miller, R. 2000. *Fat of the Land: Garbage in New York: The Last Two Hundred Years*. New York: Four Walls Eight Windows.

87. Newman, W. A. and Holton, W. E. 2006. *Boston's Back Bay: The Story of America's Greatest 19th Century Landfill Project*. Boston: Northeastern University Press.

88. Noris, R. D. and Matthews, J. E. 1994. *Handbook of Bioremediation*. Boca Raton, FL: Lewis Publishers.

89. Nyer, E. K. 1993. *Practical Techniques for Groundwater and Soil Remediation*. Boca Raton, FL: Lewis Publishers.

90. Oakes, J. M. 1997. The Location of Hazardous Waste Facilities. Dissertation, University of Massachusetts.

91. Quivik, F. L. 2001. Integrating the Preservation of Cultural Resources with Remediation of Hazardous Materials: An Assessment of Superfund's Record. *Public Historian* 23(2): 47–61.

92. Riser-Roberts, E. 1998. *Remediation of Petroleum Contaminated Soils: Biological, Physical, and Chemical Processes*. Boca Raton, FL: Lewis Publishers.

93. Rogoff, M. J. and Williams, J. F. 1994. *Approaches to Implementing Solid Waste Recycling Facilities*. Park Ridge, NJ: Noyes Publications.

94. Saha, R. and Mohai, P. 2005. Historical Context and Hazardous Waste Facility Siting: Understanding Temporal Patterns in Michigan. *Social Problems* 52(4): 618–648.

95. Sara, M. N. 2003. *Site Assessment and Remediation Handbook*. Boca Raton, FL: Lewis Publishers.

96. Sharangpani, S. 1991. The Jersey Station Landfill: A Reconstruction of Its History and an Assessment of Its Impact on Groundwater. Dissertation, Southern Illinois University, Edwardsville.

97. Strathman, J. G., et al. 1995. The Demand for Solid Waste Disposal. *Land Economics* 71(1): 57–64.

98. Suthersan, S. S. 1997. *Remediation Engineering: Design Concepts*. Boca Raton, FL: CRC-Lewis Publishers.

99. Wagner, T. 2004. Hazardous Waste: Evolution of a National Environmental Problem. *Journal of Policy History* 16(4): 306–331.

100. Williams, P. T. 2005. *Waste Treatment and Disposal*. Hoboken, NJ: Wiley.

101. Wilson, D. J. 1994. *Hazardous Waste Site Soil Remediation: Theory and Application of Innovative Technologies*. New York: Marcel Dekker.

102. Wulfhorst, J. D. 2000. Collective Identity and Hazardous Waste Management. *Rural Sociology* 65(2): 275–294.

READINGS IN HAZARDOUS WASTE MANAGEMENT IN POLLUTION PREVENTION

103. Hamilton, J. T. 1993. Politics and Social Costs: Estimating the Impact of Collective Action on Hazardous Waste Facilities. *RAND Journal of Economics* 24(1): 101.

104. Hirshfeld, S., et al. 1992. Assessing the True Costs of Landfills. *Waste Management & Research* 10(6): 471.

105. Kunreuther, H., et al. 1993. Siting Noxious Facilities: A Test of the Facility Siting Credo. *Risk Analysis* 13(3): 301.

106. Massmann, J. W. 1998. *Groundwater Contamination from Waste Management Sites: The Interaction between Risk-Based Engineering Design and Regulatory Policy.* Ottawa: National Library of Canada.

107. Qasim, S. R. and Chiang, W. W. 1994. *Sanitary Landfill Leachate: Generation, Control, and Treatment*. Lancaster, PA: Technomic Publishing.

108. Savas, E. S., ed. 1977. *The Organization and Efficiency of Solid Waste Collection*. Lexington, MA: Heath.

109. White, P., et al. 1995. *Integrated Solid Waste Management: A Lifecycle Inventory*. New York: Blackie Academic & Professional.

READINGS IN HEALTH AND SAFETY

110. An, H., et al. 1999. Occupational Health and Safety amongst Municipal Solid Waste Workers in Florida. *Waste Management Research* 17(5): 369.

111. Brown, J. E., et al. 1995. Hypersensitivity Pneumonitis from Residential Composting: Residential Composter's Lung. *Annals of Allergy Asthma Immunology* 74: 45–47.

112. Cohen, J. T., et al. 1997. Life Years Lost at Hazardous Waste Sites: Remediation Worker Fatalities vs. Cancer Deaths to Nearby Residents. *Risk Analysis* 17(4): 419–425.

113. Cordier, S., et al. 2004. Risk of Congenital Anomalies in the Vicinity of Municipal Solid Waste Incinerators. *Occupational & Environmental Medicine* 61(1): 8–15.

114. Denison, R. A. and Silbergeld, E. K. 1988. Risks of Municipal Solid Waste Incineration: An Environmental Perspective. *Risk Analysis* 8(3): 343–355.

115. El-Fadel, M., et al. 1997. Environmental Impacts of Solid Waste Land Filling. *Journal of Environmental Management* 50(1): 1–26.

116. Gaby, W. L. 1981. Health Hazards Associated with Solid Waste Disposal. *Reviews in Environmental Health* 3(3): 277–291.

117. Gellin, G. A. 1985. Dermatoses Acquired by Solid Waste Handlers. *American Journal of Industrial Medicine* 8(4–5): 363–370.

118. Gochfeld, M. 2004. Risk-Risk Balancing for Hazardous Waste Workers. *Risk Analysis* 24(2): 347–348.

119. Heida, H., et al. 1995. Occupational Exposure and Indoor Air Quality Monitoring in a Composting Facility. *AIHA Journal* 56(1): 39–43.

120. Hoskin, A. F., et al. 1994. Estimated Risk of Occupational Fatalities Associated with Hazardous Waste Site Remediation. *Risk Analysis* 14(6): 1011–1017.

121. Koplin, A. N. 1988. Septic Tanks Can Be Dangerous to Your Health. *Journal of Public Health Policy* 9(2): 164–166.

122. Krajewski, J. A., et al. 2002. Occupational Exposure to Organic Dust Associated with Municipal Waste Collection and Management. *International Journal of Occupational Medicine & Environmental Health* 15(3): 289–301.

123. Kramer, M. N., et al. 1989. Allergic Bronchopulmonary Aspergillosis from a Contaminated Dump Site. *American Review of Respiratory Disease* 140: 1086–1088.

124. Lisk, D. J. 1988. Environmental Implications of Incineration of Municipal Solid Waste and Ash Disposal. *Science of the Total Environment* 74: 39–66.

125. Lundholm, M. and Rylander, R. 1980. Occupational Symptoms among Compost Workers. *Journal of Occupational Medicine* 22: 256–257.

126. Malkin, R., et al. 1992. Blood Lead Levels in Incinerator Workers. *Environmental Research* 59(1): 265–270.

127. NIOSH. 1997. *Preventing Worker Injuries and Deaths from Moving Refuse Collection Vehicles* (Pub. 97–110). Cincinnati: National Institute for Occupational Safety and Health.

128. Poulsen, O. M., et al. 1995. Sorting and Recycling of Domestic Waste. Review of Occupational Health Problems and Their Possible Causes. *Science of the Total Environment* 168(1): 33–56.

129. Poulsen, O. M., et al. 1995. Collection of Domestic Waste: Review of Occupational Health Problems and Their Possible Causes. *Science of the Total Environment* 170(1–2): 1–19.

130. Romine, J. D. and Barth, E. F. 2002. Preliminary Assessment of Worker and Ambient Air Exposure During Soil Remediation Technology Demonstration. *AIHA Journal* 63(1): 76–80.

131. Rosenman, K. D., et al. 2003. Cleaning Products and Work-Related Asthma. *Journal of Occupational & Environmental Medicine* 45(5): 556–563.

132. Sigsgaard, T., et al. 1994. Respiratory Disorders and Atopy in Danish Refuse Workers. *American Journal of Respiratory & Critical Care Medicine* 149: 1407–1412.

133. Still, J. M., et al. 1996. Burns Due to Flammable Solvents Ignited with Floor Buffers. *Journal of Burn Care Rehabilitation* 17(2): 188–190.

Sector 61: Educational Services

6111: Elementary and Secondary Schools
6112: Junior Colleges
6113: Colleges, Universities, and Professional Schools
6114: Business Schools and Computer and Management Training
6115: Technical and Trade Schools
6116: Other Schools and Instruction
6117: Educational Support Service

READINGS IN HISTORY

1. Adams, D. P. 1980. The American Junior/Community College: Historical Roots, Contemporary Trends, and Implications for Growth. Dissertation, University of Northern Colorado.
2. Asato, N. 2003. Ousting Japanese Language Schools. *Pacific Northwest Quarterly* 94(3): 140–150.
3. Benson, C. S. and Lareau, A. P. 1982. The Uneasy Place of Vocational Education. *Education & Urban Society* 15(1): 104–124.
4. Bers, T. H. 1980. Politics, Programs, and Local Governments: The Case of Community Colleges. *Journal of Politics* 42(1): 150–164.
5. Cohen, D. K. and Grant, S. G. 1993. America's Children and Their Elementary Schools. *Daedalus* 122(1): 177–207.
6. Cooper, B. S. 1984. The Changing Demography of Private Schools: Trends and Implications. *Education & Urban Society* 16(4): 429–442.

7. Cordasco, F. 1982. Bilingual Education in American Schools: A Bibliographic Essay. *Immigration History Newsletter* 14(1): 1–8.

8. Edwards, N. J. 1982. The Public Community College in America: Its History, Present Condition, and Future Outlook with Special Reference to Finance. Dissertation, Claremont Graduate School.

9. Ennels, J. A. 2002. The "Wright Stuff": Pilot Training at America's First Civilian Flying School. *Air Power History* 49(4): 22–31.

10. Gallagher, E. A. 2001. Protecting and Strengthening the Dream: California Junior Colleges during the Great Depression. *Michigan Academician* 33(3): 231–242.

11. Gordon, E. E. 1989. Centuries of Tutoring: A Perspective on Childhood Education. Dissertation, Loyola University of Chicago.

12. Harrell, C. M. 1996. A Historical Description of Marion County, Texas, Elementary and Secondary Schools. Dissertation, East Texas State University.

13. Hechinger, F. M. 1993. Schools for Teenagers: A Historic Dilemma. *Teachers College Record* 94(3): 522–539.

14. Jacoby, D. 1996. Plumbing the Origins of American Vocationalism. *Labor History* 37(2): 235–272.

15. Johnson, K. 1987. The State of Elementary and Secondary Education in Iowa in 1900. *Annals of Iowa* 49(1–2): 26–57.

16. Kantor, H. 1986. Work, Education, and Vocational Reform: The Ideological Origins of Vocational Education, 1890-1920. *American Journal of Education* 94(4): 401–426.

17. Libbey, J. K. 2004. Flight Training and American Aviation Pioneers. *American Aviation Historical Society Journal* 49(1): 28–36.

18. Love, T. M. 1994. The Civilian Pilot Training Program at Parks Air College. *American Aviation Historical Society Journal* 39(2): 140–147.

19. Manos, M. A. 1997. Muse and Mettle: Fine Arts Instruction in American Public Schools during WW II. *JMHES* 24: 18–22.

20. Matsuda, L. G. 1987. A Historical Examination of Vocabulary Development in the Schools, 1900–1985. Dissertation, University of Hawaii.

21. McCabe, H. M. 1981. Cooperative Education in the Community College: A Critical Analysis. Dissertation, Columbia University.

22. Miller, J. A. 1980. Public Elementary Schools in Cincinnati 1870–1914. *Cincinnati Historical Society Bulletin* 38(2): 82–95.

23. Oakley, J. R. 1979. The Origins and Development of the Public Junior College Movement, 1850–1921. Dissertation, University of North Carolina, Greensboro.

24. Pedersen, R. P. 2000. The Origins and Development of the Early Public Junior College, 1900–1940. Dissertation, Columbia University.

25. Pietig, J. 1983. Values and Morality in Early 20th Century Elementary Schools: A Perspective. *Social Education* 47(4): 262–265.

26. Ratcliff, J. L. 1987. First Public Junior Colleges in an Age of Reform. *Journal of Higher Education* 58(2): 151–180.

27. Tudor, W. W. 1968. Knight School: A Report on 25 Years of Helicopter Training. *Aerospace Historian* 16(3): 9–11, 44–46.

READINGS IN HAZARDOUS WASTE MANAGEMENT AND POLLUTION PREVENTION

28. Ashbrook, P. C. and Reinhardt, P. A. 1985. Hazardous Wastes in Academia. *Environmental Science & Technology* 19(12): 1150–1155.

29. Bass, T. M. and Frick, M. J. An Examination of Pollution Prevention in Montana Secondary Agricultural Education Laboratories. In *Proceedings of the 27th Annual National Agricultural Education Research Conference*, 129–143.

30. Budny, M. H. 1998. Hazardous and Special Waste Pollution Prevention Opportunity Assessment for a Metropolitan School District. Dissertation, University of Nebraska at Lincoln.

31. Hageman, J. P. 2002. Handling, Storage, Treatment, and Disposal of Mixed Wastes at Medical Facilities and Academic Institutions. *Health Physics* 82(5): S66–S76.

32. Kaufman, J. A. 1990. *Waste Disposal in Academic Institutions*. Chelsea, MI: Lewis Publishers.

33. Li, E., et al. 2003. Pollution Prevention Guideline for Academic Laboratories. *Journal of Chemical Education* 80(1): 45–49.

READINGS IN HEALTH AND SAFETY

34. Boden, B. P., et al. 2003. Catastrophic Cheerleading Injuries. *American Journal of Sports Medicine* 31(6): 881–888.

35. Bowling, A. 1989. Injuries to Dancers: Prevalence, Treatment, and Perceptions of Cause. *BMJ* 298(6675): 731–734.

36. Daisey, J. M., et al. 2003. Indoor Air Quality, Ventilation and Health Symptoms in Schools: An Analysis of Existing Information. *Indoor Air* 13(1): 53–64.

37. Dougherty, J. D. 1969. Variation in the Annual General Aviation Accident Rate As It Relates to Flight Training. *Aerospace Medicine* 40(7): 699–707.

38. Guidotti, S., et al. 1982. Multiple Myeloma in Cosmetologists. *American Journal of Industrial Medicine* 3(2): 169–171.

39. Jaffe, L. and Minkoff, J. 1988. Martial Arts: A Perspective on Their Evolution, Injuries, and Training Formats. *Orthopedic Reviews* 17(2): 208–221.

40. Melton, C. E., et al. 1975. Effect of a General Aviation Trainer on the Stress of Flight Training. *Aviation Space Environmental Medicine* 46(1): 1–5.

41. Palmer, A., et al. 1979. Respiratory Disease Prevalence in Cosmetologists and Its Relationship to Aerosol Sprays. *Environmental Research* 19(1): 136–153.

42. Reutzel, T. J., et al. 2001. Medication Management in Primary and Secondary Schools. *Journal of the American Pharmaceutical Association* 41(1): 67–77.

43. Santilli, J. 2002. Health Effects of Mold Exposure in Public Schools. *Current Allergy & Asthma Report* 2(6): 460–467.

44. Teitz, C. C. 1982. Sports Medicine Concerns in Dance and Gymnastics. *Pediatric Clinics of North America* 29(6): 1399–1421.

45. Teta, M. J., et al. 1984. Cancer Incidence among Cosmetologists. *Journal of the National Cancer Institute* 72: 1051–1057.

46. Uhoda, E., et al. 2005. The Conundrum of Skin Pores in Dermocosmetology. *Dermatology* 210(1): 3–7.

Sector 62: Health Care and Social Assistance

621: Ambulatory Health Care Services
6211: Offices of Physicians
6212: Offices of Dentists
6213: Offices of Other Health Practitioners
6214: Outpatient Care Centers
6215: Medical and Diagnostic Laboratories
6216: Home Health Care Services
6219: Other Ambulatory Health Care Services

READINGS IN HISTORY

1. Anonymous. 1981. The Levy Brothers: Early San Francisco Chiropodists: A Picture Story. *Western States Jewish Historical Quarterly* 13(4): 317–321.

2. Carfaro, J. A. 1998. From Personal Tragedy to Public Problem: Miracle Technologies and End-Stage Renal Disease Policy, 1960–1972. Dissertation, Case Western Reserve University.

3. Charles, A. D. 1981. History of Dentistry in South Carolina. Dissertation, University of South Carolina.

4. Connor, J. T. H. 1994. Dutch Technological Migration and North American Commercial Exploitation: Dr. Willem Kolff and the Development of the Artificial Kidney. *European Contributions to American Studies* 31: 281–303.

5. Cremin, J. H. 1985. The Los Angeles Clinical Laboratory Medical Group: The Early Years. *Southern California Quarterly* 67(4): 389–444.

6. Curtis, E. K. 1994. The French Revolution and the American Invention of Dentistry. *Journal of American Culture* 17(3): 13–16.

7. Daniels, R. 1991. Legislation and the American Dialysis Industry: Some Considerations about Monopoly Power in Renal Care. *American Journal of Economics & Sociology* 50(2): 223–242.

8. Fales, M. J. H. 1975. History of Dental Hygiene Education in the U.S., 1913–1975. Dissertation, University of Michigan.

9. Fraundorf, K. C. 1975. Competition in Blood Banking. *Public Policy* 23(2): 219–240.

10. Gibbons, R. W. 1981. Physician-chiropractors: Medical Presence in the Evolution of Chiropractic. *Bulletin of the History of Medicine* 55(2): 233–245.

11. Given, R. S. 1995. An Economic Analysis of the California HMO Industry, 1986–1992. Dissertation, University of California, Berkeley.

12. Haas-Wilson, D. 1986. The Effect of Commercial Practice Restrictions: The Case of Optometry. *Journal of Law & Economics* 29(1): 165–186.

13. Hiott, S. G. 1990. Osteopathy in South Carolina: The Struggle for Recognition. *South Carolina Historical Magazine* 91(3): 195–209.

14. Horn, A. T. 1985. The Professionalization of Ambulance Services. Dissertation, University of Michigan.

15. Hunt, E. E. 2003. Oral History: How Americans Got Their Straight White Teeth. Dissertation, University of Pennsylvania.

16. Karow, P. J. 1991. The Professionalization of Physical Therapy in the U.S. Dissertation, New York University.

17. Kennedy, L. 1978. Community Blood Banking in the U.S. from 1937–1975: Organizing Formation, Transformation and Reform in a Climate of Competing Ideologies. Dissertation, New York University.

18. Levitt, J. 1979. Politics, Technology and Labor in the Clinical Laboratories: A Study in the Reinforcement of Private Power and Hierarchical Control. Dissertation, New York University.

19. Martin, S. C. 1994. The Only True Scientific Method of Healing: Chiropractic and American Science, 1895–1990. *Isis* 85(2): 206–227.

20. Maurizi, A. R., et al. 1981. Competing for Professional Control: Professional Mix in the Eyeglass Industry. *Journal of Law & Economics* 24(2): 351–364.

21. Morman, E. T. 1984. Clinical Pathology in America, 1865–1915: Philadelphia as a Test Case. *Bulletin of the History of Medicine* 58(2): 198–215.

22. Penny, A. O. 1968. Nehemiah Kenison: New England's Pioneer Podiatrist. *New England Galaxy* 9(4): 46–52.

23. Picard, A. 2004. Making the American Mouth: Dental Professionalism, Dental Public Health, and the Construction of Identity in the 20th Century United States. Dissertation, University of Michigan.

24. Richardson, J. K. 1983. A Study of the Forces and Influences Leading to the Development of Specialization in the Profession of Physical Therapy in the U.S. Dissertation, University of Pittsburgh.

25. Ring, M. E. 1995. Behind the Dentist's Drill. *American Heritage of Invention & Technology* 11(2): 24–31.

26. Schmitt, R. C. 1983. Royal Crowns and Historic Plaque: Dentistry in Hawaii During the 19th Century. *Hawaiian Journal of History* 17: 143–155.

27. Smith-Cunnien, S. 2005. Without Drugs or Knives: The Early Years of Chiropractic. *Minnesota History* 59(5): 200–212.

READINGS IN HAZARDOUS WASTE MANAGEMENT AND POLLUTION PREVENTION

28. Albuquerque Public Works Department. 1999. *Pollution Prevention Guidelines for Dental Offices.* Albuquerque: Public Works Department.

29. Boatright, D.T., et al. 1995. A Comprehensive Biomedical Waste Survey. *Journal of Environmental Health* 57(8): 15.

30. Davies, T. and Lowe, A. L. 1999. *Environmental Implications of the Health Care Service Sector.* Washington, DC: Resources for the Future.

31. New Hampshire Department of Environmental Services. 2002. *Best Management Practices for Dental Offices in New Hampshire* (NHDES-WMD-2001). Concord, NH: Pollution Prevention Program.

32. U.S. Environmental Protection Agency. 2005. *Profile of the Healthcare Industry* (EPA/310/R-05/002). Washington, DC: Office of Compliance.

READINGS IN HEALTH AND SAFETY

33. Alvord, L. S. 1990. Noise Levels in Physician's Offices. *Ear Hear* 11(5): 391–393.

34. Anonymous. 1991. Hemodialysis Machines. *Health Devices* 20(6): 187–232.

35. Anonymous. 1992. OSHA Workplace Requirements in Effect: Direct Impact on Physician Offices. *Colorado Medicine* 89(3): 76–77.

36. Cedar, W. R. 1973. Mercury Vapor: A Hazard in the Dentist's Office. *Dental Studies* 51(5): 28–30.

37. Crandall, C. E. 1959. Radiation in Dental Offices. *Journal of the American Dental Association* 59: 971–975.

38. Curry, C. J. 1957. Traffic Accidents and Safety: Transportation of the Injured. *Journal of the Michigan Medical Society* 56(9): 1139.

39. Erich, J. 2002. Ambulance Safety: What's New, What's Needed. *Emergency Medical Services* 31(6): 51–54.

40. Evans, M. W., et al. 2007. Assessment and Risk Reduction of Infectious Pathogens on Chiropractic Treatment Tables. *Chiropractic Osteopat* 15(1): 8.

41. Feldman, T. B., et al. 1997. Violence in Medical Facilities: A Review of 40 Incidents. *Journal of the Kentucky Medical Association* 95(5): 183–189.

42. Hales, R. H. 1970. Ocular Injuries Sustained in the Dental Office. *American Journal of Ophthalmology* 70(2): 221–223.

43. Harrington, J. M. 1982. Health and Safety in Medical Laboratories. *Bulletin of the World Health Organization* 60(1): 9–16.

44. Herwaldt, L. A., et al. 1998. Infection Control in the Outpatient Setting. *Infection. Control & Hospital Epidemiology* 19(1): 41–74.

45. Jacob, J. A. 1995. Dental Office Safety. *CDS Review* 88(2): 14–20.

46. Kliger, A. S. 2007. The Dialysis Medical Director's Role in Quality and Safety. *Seminars in Dialysis* 20(3): 261–264.

47. Nordberg, M. 2006. Improving Ambulance Safety. *Emergence Medical Services* 35(6): 78.

48. Roggenkamp, C. L. and Tune, T. E. 1982. Relationship of Mercury Vapor Levels in a Dentist's Clinic, Shoes and Home: A Case Report. *Journal of the Indiana Dental Association* 61(2): 23–26.

49. Thomas, G. W. 1957. Safety Practices in the Clinical Laboratory. *American Journal of Medical Technology* 23(3): 159–162.

622: Hospitals
6221: General Medical and Surgical Hospitals
6222: Psychiatric and Substance Abuse Hospitals
6223: Specialty (Except Psychiatric and Substance Abuse) Hospitals

READINGS IN HISTORY

50. Achenbaum, W. A. 1985. American Medical History, Social History, and Medical Policy. *Journal of Social History* 18(3): 343–348.

51. Albert, M. R. 2000. A Dermatology Ward at the Beginning of the 20th Century. *Journal of the American Academy of Dermatology* 42(1): 113–123.

52. Baker, J. P. 1996. *The Machine in the Nursery: Incubator Technology and the Origins of Newborn Intensive Care*. Baltimore, MD: Johns Hopkins University Press.

53. Brush, B. L. 2001. Historical Analysis of Side Rail Use in American Hospitals. *Journal of Nursing Scholarship* 33(4): 381–385.

54. Burns, C. R. 1993. The Development of Galveston's Hospitals during the 19th Century. *Southwestern Historical Quarterly* 97(2): 238–263.

55. Clevenger, S. 2001. St. Vincent's and the Sisters of Providence: Oregon's First Permanent Hospital. *Oregon Historical Quarterly* 102(2): 210–221.

56. Connor, J. T. H. 1990. Hospital History in Canada and the U.S. *Canadian Bulletin of Medical History* 7(1): 93–104.

57. Creighton, C. 1993. Graded Activity: Legacy of the Sanatorium. *American Journal of Occupational Therapy* 47(8): 745–748.

58. Cruse, J. M. 1999. History of Medicine: The Metamorphosis of Scientific Medicine in the Ever Present Past. *American Journal of Medical Science* 318(3): 171–180.

59. Garrett, V. 1996. Substance Abuse Treatment in Southern California: The History and Significance of the Antelope Valley Rehabilitation Centers. *Journal of Policy History* 8: 181–205.

60. Geller, J. L. 1994. Issues in American Psychiatry Reflected in the Remarks of APA Presidents, 1844–1994. *Hospital Comm. Psych.* 45(10): 993–1004.

61. Luckey, V. M. 1969. The First Rural Public Hospital. *Annals of Iowa* 40(1): 68–72.

62. Lynaugh, J. E. 1996. *American Nursing: From Hospitals to Health Systems.* Cambridge: Blackwell.

63. Mayer, S. 1998. Inclusion and Reform: The Contemporary History of a State Mental Hospital. Dissertation, Northwestern University.

64. Morantz-Sanchez, R. 1988. The Triumph and Tragedy of the American Hospital. *Reviews in American History* 16(4): 526–531.

65. Peltin, N. 2004. *We Stand By to Assist You*. Seattle: Swedish Medical Center.

66. Risse, G. B. 1999. *Mending Bodies, Saving Souls: A History of Hospitals*. New York: Oxford University Press.

67. Schell, E. 1993. The Origins of Geriatric Nursing. *Nursing History Review* 1: 203–216.

68. Stevens, R. 1999. *In Sickness and in Wealth: American Hospitals in the 20th Century*. Baltimore, MD: Johns Hopkins University Press.

69. Vilensky, J. A., et al. 2004. Denny-Brown, Boston City Hospital, and the History of American Neurology. *Perspectives in Biology & Medicine* 47(4): 505–518.

70. Wall, B. M. 2003. Science and Ritual: The Hospital as Medical and Sacred Space, 1865–1920. *Nursing History Review* 11: 51–68.

71. Warner, J. H. and Tighe, J. A., eds. 2001. *Major Problems in the History of American Medicine and Public Health: Documents and Essays*. Boston: Houghton Mifflin.

72. Wengert, J. W. 1997. The Contract Surgeon. *Journal of the West* 36(1): 67–76.

73. Willis, F. N. 1981. The First Mental Hospital for American Indians, 1900–1934. *Bulletin Menninger Clinic* 45: 149–154.

74. Zetka, J. R. 2003. *Surgeons and the Scope*. Ithaca, NY: Cornell University Press.

READINGS IN HAZARDOUS WASTE MANAGEMENT AND POLLUTION PREVENTION

75. Acharya, D. B. and Singh, M. 2000. *The Book of Hospital Waste Management*. New Delhi: Minvera Press.

76. Barker, J. P. and Abdelatti, M. O. 1997. Anesthetic Pollution. Potential Sources, Their Identification and Control. *Anaesthesia* 52(11): 1077–1083.

77. Burd, M. 2005. Reducing the Risks Related to the Handling and Disposal of Health Care Waste. *Professional Nurse* 20(8): 40–42.

78. Chitnis, V., et al. 2005. Biomedical Waste in Laboratory Medicine: Audit and Management. *Indian Journal of Medical Microbiology* 23(1): 6–13.

79. City of Palo Alto. n.d. *Pollution Prevention for Hospitals & Medical Facilities*. http://www.c2p2online.com/documents/P2forhospitals.pdf (accessed 7 November 2008).

80. Daschner, F. D. and Dettenkofer, M. 1997. Protecting the Patient and the Environment: New Aspects and Challenges in Hospital Infection Control. *Journal of Hospital Infection* 36(1): 7–15.

81. Dharan, S. and Pittet, D. 2002. Environmental Controls in Operating Theatres. *Journal of Hospital Infection*. 51(2): 79–84.

82. Dilly, G. A. and Shanklin, C. W. 2000. Solid Waste Management Practice in U.S. Army Medical Treatment Facilities. *Military Medicine* 165(4): 302–304.

83. Eitel, D. F. 1998. Managing Medical Waste Technology: How U.S. Hospitals Adapted to Change. *International Journal of Technology Management* 15(3/5): 402–420.

84. English, N. 1996. Reprocessing and Disposables: One Strategy to Balance Cost Reduction and Quality Patient Care. *Today's Surgical Nurse* 18(4): 23–26.

85. McNally, R. D. 2006. Acquiring New Technology and Surviving Environmental Pressures. *Radiological Technology* 77(3): 191–199.

86. Melamed, A. 2003. Environmental Accountability in Perioperative Settings. *AORN Journal* 77(6): 1157–1168.

87. New Hampshire Department of Environmental Services. 2003. *Pollution Prevention Resources for Healthcare Facilities* (NHDES-WMD-03–1). http://des.nh.gov/organization/commissioner/pip/publications/wmd/documents/nhdes-wmd-03-1.pdf (accessed 7 November 2008).

88. Phillips, G. 1999. Microbiological Aspects of Clinical Waste. *Journal of Hospital Infection*. 41(1): 1–6.

89. Rutala, W. A., et al. 1989. Management of Infectious Waste by US Hospitals. *JAMA* 262(12): 26–29.

90. Scottish Environmental Protection Agency. 2004. *Pollution Prevention Guidelines: Hospitals & Health Establishments* (PPG-25). Stirling, UK: SEPA.

91. Thornton, J., et al. 1996. Hospitals and Plastics. Dioxin Prevention and Medical Waste Incinerators. *Public Health Reports* 111(4): 298–313.

92. Topf, M. 2000. Hospital Noise Pollution: An Environmental Stress Model to Guide Research and Clinical Interventions. *Journal of Advances in Nursing* 31(3): 520–528.

93. Wong, K.-F. V., et al. 1994. Medical Waste Characterization. *Journal of Environmental Health* 57(1): 19.

94. Zemelman, M. S. 1996. Environmental Compliance for Hospitals: A Practical Guide. *Journal of Health & Hospital Law* 29(1): 7–13.

READINGS IN HEALTH AND SAFETY

95. Aghababian, R., et al. 1994. Disasters within Hospitals. *Annals of Emergency Medicine* 23(4): 771–777.

96. Fontaine, D. K. and Gerardi, D. 2005. Healthier Hospitals? *Nursing Management* 36(10): 34–44.

97. Gardner, V. F. 1971. NFPA Hospital Safety Standards. *Hospitals* 45(17): 102–105.

98. Kane-Gill, S. and Weber, R. J. 2006. Principles and Practices of Medication Safety in the ICU. *Critical Care Clinics* 22(2): 273–290.

99. Lohman, J. 1982. Hospital Safety Strategies. *Occupational Health & Safety* 51(12): 46–47.

100. Scott, L. F. 1956. Why Emphasize Hospital Safety? *Hospital Management* 81(2): 45–46.

101. Triolo, P. K. 1989. Occupational Health Hazards of Hospital Staff Nurses. *AAOHN Journal* 37(7): 274–279.

102. Unruh, L., et al. 2005. Managed Care and Patient Safety: Risks and Opportunities. *Health Care Management* 24(3): 245–256.

103. Wholey, D., et al. 2004. The Environmental Context of Patient Safety and Medical Errors. *Journal of Rural Health* 20(4): 304–313.

623: Nursing and Residential Care Facilities
6231: Nursing Care Facilities
6232: Residential Mental Retardation, Mental Health, and Substance Abuse Facilities
6233: Community Care Facilities for the Elderly
6239: Other Residential Care Facilities

READINGS IN HISTORY

104. Binney, E. A. 1995. The Transformation of Community-Based Long-term Care, 1981–1991. Dissertation, University of California, San Francisco.

105. Buhler-Wilkerson, D. K. 2001. *No Place Like Home: A History of Nursing and Home Care in the United States.* Baltimore, MD: Johns Hopkins University Press.

106. Christopherson, V. A. 1972. Retirement Communities: The Cities of Two Tales. *Social Science* 47(2): 82–86.

107. Eagle, E. 1968. Nursing Homes and Related Facilities: A Review of the Literature. *Public Health Reports* 83(8): 673–684.

108. Felton, G. S. 1969. Psychosocial Aspects of Names of Retirement Facilities. *Names* 17(4): 284–292.

109. Hovey, W. 2000. The Worst of Both Worlds: Nursing Home Regulation in the United States. *Policy Studies Review* 17(4): 43–59.

110. Hynes, C. J. 1980. The Regulation of Nursing Homes: A Case Study. *Proceedings of the Academy of Political Science* 33(4): 126–136.

111. Mendelson, M. A. and Hapgood, D. 1974. The Political Economy of Nursing Homes. *Annals of the American Academy of Political and Social Science* 415: 95–105.

112. Schell, E. 1993. The Origins of Geriatric Nursing: The Chronically Ill Elderly in Almhouses and Nursing Homes, 1900–1950. *Nursing History Review* 1: 203–216.

113. Williams, T. 1977. An Analysis of Life Histories of the Elderly in a Retirement Home. *Kroeber Anthropological Society Papers* 49: 65–74.

114. Young, T. M., et al. 1988. Residential Group Care for Children Considered Emotionally Disturbed, 1966–1981. *Social Service Review* 62(1): 158–170.

READINGS IN HEALTH AND SAFETY

115. Crossley, K., et al. 1990. Needlestick Injuries and Needle Disposal in Minnesota Nursing Homes. *Journal of the American Geriatric Society* 38(7): 793–796.

116. Hawes, C., et al. 2005. Assisted Living in Rural America: Results from a National Survey. *Journal of Rural Health* 21(2): 131–139.

117. Hodder, S. L., et al. 1995. Acute Respiratory Illness in Older Community Residents. *Journal of the American Geriatric Society* 43(1): 24–29.

118. Hotchkiss, B. L. 1953. A Report from California: Progress in Nursing, Convalescent and Rest Homes. *Hospitals* 27(1): 75–76.

119. Hovey, W. 2000. The Worst of Both Worlds: Nursing Home Regulation in the United States. *Policy Study Reviews* 17(4): 43–59.

120. Landy, D. and Greenblatt, M. 1965. Study of a Halfway House. *Rehabilitation Record* 52: 14–15.

121. Manton, K. G. 2003. Forecasting the Nursing Home Population. *Medical Care* 41(1): 21–24.

122. Mitty, E. L. 2003. Policy Perspectives: Assisted Living and the Role of Nursing. *American Journal of Nursing* 103(8): 32–43.

123. Mollica, R. L. 1998. Managed Care and Assisted Living: Trends and Future Pprospects. *Journal of Health & Human Services Administration* 20(3): 264–280.

124. Pinel, C. 1988. The Safety of Patients in Geriatric Hospitals. *Nursing* 3(30): 20–23.

125. Rubington, E. 1967. The Halfway House for the Alcoholic. *Mental Hygiene* 51(4): 552–560.

126. Scott-Cawiezell, J., et al. 2006. Nursing Home Safety: A Review of the Literature. *Annual Review of Nursing Research* 24: 179–215.

127. ——, et al. 2007. Nursing Home Error and Level of Staff Credentials. *Clinical Nursing Research* 16(1): 72–78.

128. Wallace, M. 2003. Is There a Nurse in the House? The Role of Nurses in Assisted Living: Past, Present and Future. *Geriatric Nursing* 24(4): 218–221.

129. Ward, D. J. 2001. Infection Control Policies in Old Folks Homes. *Nursing Standard* 15(46): 40–44.

624: Social Assistance
6241: Individual and Family Services
6242: Community Food and Housing, and Emergency and Other Relief Services
6243: Vocational Rehabilitation Services
6244: Child Day Care Services

READINGS IN HISTORY

130. Aguilar, G. D. 1986. Development of Domestic Violence Shelters as Social Service Delivery Systems. Dissertation, University of California, Berkeley.

131. Berkowitz, E. D. 1994. A Historical Preface to the Americans with Disabilities Act. *Journal of Policy History* 6(1): 96–119.

132. Best, R. H. 1976. A History of the Colorado Vocational Rehabilitation Program. Dissertation, University of Northern California.

133. Brown, L. D. 1975. A History of the Vocational Rehabilitation of Veterans with Psychiatric Disabilities. Dissertation, University of Northern Colorado.

134. Cohen, A. J. 1996. A Brief History of Federal Financing for Child Care in the U.S. *Future of Children* 6(2): 26–40.

135. Davis, W.L. 1991. Family Planning Services: A History of U.S. Federal Legislation. *Journal of Modern History* 16(4): 381–400.

136. Driver, F. 1993. *Power and Pauperism: The Workhouse System, 1834–1884.* New York: Cambridge University Press.

137. Frankel, H. 1988. Family-centered, Home-based Services in Child Protection: A Review of the Research. *Social Science* Review 62(1): 137–157.

138. Goldmintz, L. R. 1987. The Growth of Day Care Services, 1890–1946. Dissertation, Yeshiva University.

139. Gounis, K. N. 1993. The Domestication of Homelessness: The Politics of Space and Time in New York City Shelters. Dissertation, Columbia University.

140. Hanson, C. S. 1992. The History of Work in Physical Dysfunction. *American Journal of Occupational Therapy* 46(1): 56–62.

141. Haskins, R. 1979. Day Care and Public Policy. *Urban and Social Change Review* 12(1): 3–10.

142. Hoover, J. A. 1996. Diversional Occupational Therapy in WWI: A Need for Purpose in Occupations. *American Journal of Occupational Therapy* 50(10): 881–885.

143. Radis, C. M. 2003. The History of the Day Nursery (1870s–1930s) and Its Relevance to Child Day Care Today. Dissertation, New School University.

144. Rhodes, S. L. 1979. Trends in Child Development Research Important to Day-care Policy. *Social Service Review* 53(2): 285–294.

145. Rose, E. 1999. *A Mother's Job: The History of Day Care, 1890–1960.* New York: Oxford University Press.

146. Scarr, S. 1998. American Child Care Today. *American Psychologist* 53(2): 95–108.

READINGS IN HEALTH AND SAFETY

147. Bradley, D. 2007. Safety of Young Children in Registered Day Care. *Journal of Family Health Care* 17(2): 41–42.

148. Davies, L., et al. 1981. Longitudinal Study on Elderly Recipients of Meals-on-wheels. *Journal of Human Nutrition* 35(6): 442–447.

149. Goldsmith, C. 1959. Child Day Care, a Public Health Responsibility. *American Journal of Public Health* 49: 1069–1073.

150. Jans, P. 1955. Meals on Wheels. *Nursing Outlook* 3(3): 130–131.

151. Moran, M. B. 2004. Challenges in the Meals on Wheels Program. *Journal of the American Dietetic Association* 104(8): 1219–1221.

152. Rauschenbach, B. S., et al. 1990. Dependency on Soup Kitchens in Urban Areas of New York State. *American Journal of Public Health* 80(1): 57–60.

153. Simms, M. D. 1998. Medical Care of Children Who Are Homeless or in Foster Care. *Current Opinion in Pediatrics* 10(5): 486–490.

Sector 71: Arts, Entertainment, and Recreation

711: Performing Arts, Spectator Sports, and Related Industries
7111: Performing Arts Companies
7112: Spectator Sports
7113: Promoters for Entertainment Events
7114: Agents and Managers for Artists, Athletes, Entertainers, and Other Public Figures
7115: Independent Artists, Writers, and Performers

READINGS IN HISTORY

1. Bernstein, M. F. 2001. *Football: The Ivy League Origins of an American Obsession*. Philadelphia: University of Pennsylvania Press.

2. Bialaszewski, D. W. 1980. An Economic Analysis of the Demand for Football. Dissertation, State University of New York, Buffalo.

3. Blau, J. R. and Hall, R. H. 1986. The Supply of Performing Arts in Metropolitan Places. *Urban Affairs Quarterly* 22(1): 42–65.

4. Brier, Z. 1999. Rhode Island Jews in the Performing Arts. *Rhode Island Jewish Historical Notes* 13(1): 85–92.

5. Carroll, J. M. 1999. *Red Grange and the Rise of Modern Football*. Urbana: University of Illinois Press.

6. Charlston, J. A. 2002. Disorganized and Quasi-official but Eventually Successful: Sport in the U.S. Military, 1814–1914. *International Journal of the History of Sport* 19(4): 70–88.

7. Clements, W. M. 1996. *Native American Verbal Art: Texts and Contexts*. Tucson: University of Arizona Press.

8. Daniel, P. 2004. In Their Own Words: NASCAR in American through Oral History. *Atlanta History* 46(2): 4–13.

9. Dyreson, M. 2002. Reading Football History: New Vistas in the Landscape of American Sport. *Journal of Sport History* 29(2): 202–220.

10. Evans, C. J. 2001. Applying Historical Research to Modern Orchestral Performance. Dissertation, University of Houston.

11. Fox, S. 1998. *Big Leagues: Professional Baseball, Football and Basketball in National Memory*. Lincoln: University of Nebraska Press.

12. Franks, J. 1989. California and the Rise of Spectator Sports, 1850–1900. *Southern California Quarterly* 71(4): 287–310.

13. Golden, M. 1987. Why the Square? John D. Rockefeller 3rd and the Creation of Lincoln Center for the Performing Arts. *Journal of Popular Culture* 21(3): 17–30.

14. Hall, R. L. 2002. Before NASCAR: The Corporate and Civic Promotion of Automobile Racing in the American South, 1903–1927. *Journal of Southern History* 68(3): 629–668.

15. Johnson, A. T. 1998. Minor League Baseball: Risks and Potential Benefits for Communities Large and Small. *Policy Studies Review* 15(1): 45–54.

16. Lentz, T. P. 2001. The History of the Michigan Opera Theatre. Dissertation, Wayne State University.

17. Levine, J. S. 1993. Performing Arts Patronage in New York City, 1989–1991. Dissertation, New York University.

18. Lynes, R. 1985. *The Lively Audience: A Social History of the Visual & Performing Arts in America, 1890–1950*. New York: Harper & Row.

19. Oaklander, C. I. 1999. Clara Davidge and Henry Fitch Taylor: Pioneering Promoters and Creators of American Modernist Art. Dissertation, University of Delaware.

20. Parisi, B. 1991. The History of Brooklyn's Three Major Performing Arts Institutions. Dissertation, New York University.

21. Poole, E. 2004. For the Fun of the Game: Sandlot and Semi-pro Football. *Western Pennsylvania History* 87(4): 26–32.

22. Riess, S. A. 1980. *Touching Base: Professional Baseball and American Culture in the Progressive Era*. Urbana: University of Illinois Press.

23. ——. 1994. From Pitch to Putt: Sport and Class in Anglo-American Sport. *Journal of Sport History* 21(2): 138–184.

24. Salisbury, R. H. 1988. Organized Sports and Urban Life. *Urban Affairs Quarterly* 24(2): 327–332.

25. Schultz, B. 2003. A Geographical Study of the American Ballpark. *International Journal of the History of Sport* 20(1): 126–142.

26. Seifried, C. S. 2005. An Analysis of the American Outdoor Sport Facility: Developing an Ideal Type of the Evolution of Professional Baseball and Football Structures. Dissertation, Ohio State University.

27. Taublieb, D. 2005. Arthur Weisberg's Contemporary Chamber Ensemble, 1960–1983. Dissertation, City University of New York.

28. Unsworth, A. E. 2001. Organ Pedagogy and Performance Practice in Boston, 1850–1900. Dissertation, Duke University.

29. Watterson, J. S. 2000. *College Football: History, Spectacle and Controversy.* Baltimore, MD: Johns Hopkins University Press.

30. Ziemba, J. 1999. *When Football Was Football: The Chicago Cardinals and the Birth of the NFL.* Chicago: Triumph.

READINGS IN HEALTH AND SAFETY

31. Adams, B. B. 2002. Dermatologic Disorders of the Athlete. *Sports Medicine* 32(5): 309–321.

32. Bartlett, R. M. 1997. Current Issues in the Mechanics of Athletic Activities: A Position Paper. *Journal of Biomechanics* 30(5): 477–486.

33. Bronner, S. and Brownstein, B. 1997. Profile of Dance Injuries in a Broadway Show: A Discussion of Issues in Dance Medicine Epidemiology. *Journal of Orthopedic Sports Physical Therapy* 26(2): 87–94.

34. Crocco, T. J., et al. 2004. Mathematical Determination of External Defibrillators Needed at Mass Gatherings. *Prehospital Emergency Care* 8(3): 292–297.

35. Kadel, N. J., et al. 1992. Stress Fractures in Ballet Dancers. *American Journal of Sports Medicine* 20(4): 445–449.

36. Kujala, U. M., et al. 1988. Equipment, Drugs and Problems of the Competition and Team Physician. *Sports Medicine* 6(4): 197–209.

37. Lehman, L. B. 1989. Let the Sports Spectator Beware. *Postgraduate Medicine* 85(8): 140, 142–143.

38. Perry, N. 1996. Trauma in the Stands. Fan Violence Scars Spectator Events. *Emergency Medical Services* 25(6): 50–53.

39. Scott, R. A. 2000. What Kind of Spectator Are You? *Clinical Nurse Specialist* 14(3): 105–106.

712: Museums, Historical Sites, and Like Institutions

READINGS IN HISTORY

40. Adams, W. H. 2003. Dating Historical Sites. *Historical Archeology* 37(2): 38–64.

41. Archibald, R. R. 2004. *The New Town Square: Museums and Communities in Transition*. Walnut Creek, CA: AltaMira.

42. Asma, S. T. 2001. *Soughed Animals and Pickled Heads: The Culture and Evolution of Natural History Museums.* New York: Oxford University Press.

43. Cameron, C. M. and Gatewood, J. B. 2000. Excursions into the Un-remembered Past: What People Want from Visits to Historical Sites. *Public Historian* 22(3): 107–127.

44. Gardner, J. B. 2004. Contested Terrain: History, Museums and the Public. *Public Historian* 26(4): 11–21.

45. Haines, A. L. 1981. *Historic Sites Along the Oregon Trail.* Gerald, MO: Patrice.

46. Kantor, S. G. 2002. *Alfred H. Barr, Jr. and the Intellectual Origins of the Museum of Modern Art*. Cambridge, MA: MIT Press.

47. Leibold, C. 2005. Two Hundred Years and Counting: The Pennsylvania Academy of Fine Arts. *Pennsylvania Heritage* 31(2): 6–15.

48. Mackintosh, B. 1987. The National Park Service Moves into Historical Interpretation. *Public Historian* 9(2): 50–63.

49. Magelssen, S. 2004. Living History Museums and the Construction of the Real through Performance. *Theatre Survey* 45(1): 61–74.

50. Pisney, R. F. 1977. *Historical Markers: A Bibliography*. Verona, VA: McClure.

51. Pitcaithley, D. T. 1987. Historic Sites: What Can Be Learned from Them? *History Teacher* 20(2): 207–219.

52. Reap, J. K. 1979. Legal Aspects of Historic Preservation. *Georgia Historical Quarterly* 63(1): 68–76.

READINGS IN HEALTH AND SAFETY

53. Lewis, E., et al. 2005. Disseminating a Sun Safety Program to Zoological Parks: The Effect of Tailoring. *Health Psychology* 24(5): 456–462.

54. Makos, K. A. 2001. Collections-based Health Hazards in Museums, Universities, and Cultural Institutions. *Occupational Medicine* 16(4): 649–658.

713: Amusement, Gambling, and Recreation Industries
7131: Amusement Parks and Arcades
7132: Gambling Industries
7139: Other Amusement and Recreation Industries

READINGS IN HISTORY

55. Adams, J. A. 1991. *The American Amusement Park Industry: A History of Technology & Thrills*. Boston: Twayne.

56. Adams, R. L. A. and Rooney, J. F. 1985. Evolution of American Golf Facilities. *Geographical Review* 75(4): 419–438.

57. Allen, J. B. 1983. Winter Culture: The Origins of Skiing in the United States. *Journal of American Culture* 6(1): 65–68.

58. ——. 1985. Values and Sport: The Development of New England Skiing, 1870–1940. *Oral History Review* 13: 55–76.

59. ——. 1992. The Land of Promise: American Skiing, 1840–1940. *Stadion* 18(1): 90–105.

60. ——. 1993. *From Ski Sport to Skiing: One Hundred Years of an American Sport, 1840–1940*. Amherst: University of Massachusetts Press.

61. Anderson, D. 1999. Sign, Space and Story: Roller Coasters and the Evolution of a Thrill. *Journal of Popular Culture* 33(2): 1–22.

62. Arave, J. 2002. The Forest Service Takes to the Slopes. *Utah Historical Quarterly* 70(4): 341–355.

63. Boje, D. M. 1995. Stories of the Storytelling Organization: A Postmodern Analysis of Disney as "Tamara-land." *Academy of Management Journal* 38(4): 997–1035.

64. Coleman, A. G. 1997. Culture, Landscape, and the Making of the Colorado Ski Industry. Dissertation, University of Colorado, Boulder.

65. Crang, M. 1996. Living History: Magic Kingdoms or a Quixotic Quest for Authenticity? *Tourism Research* 23(2): 415–431.

66. David, F. N. 1998. *Games, Gods and Gambling: A History of Probability and Statistical Ideas*. London: Charles Griffin & Co.

67. DeAngelis, M. 1997. Orchestrated Disorientation: Roller Coasters, Theme Parks, and Postmodernism. *Cultural Critique* 37: 107–129.

68. Dorst, J. 1993. Miniaturizing Monumentality: Theme Park Images of the American West and Confusions of Cultural Influence. *European Contributions to American Studies* 25: 254–270.

69. Eadington, W. R. 1984. The Casino Gambling Industry: A Study of Political Economy. *Annals of the American Academy of Political & Social Science* 474: 23–35.

70. Edwards, J. E. 1992. The Americanization of Nevada Gambling. *Halcyon* 14: 201–216.

71. ——. 1994. Nevada Gambling: Just Another Business Enterprise. *Nevada Historical Quarterly* 37(2): 101–114.

72. Ellis, D. 1984. Video Arcades, Youth and Trouble. *Youth & Society* 16(1): 47–65.

73. Engen, A. K. and Thompson, G. C. 2001. *First Tracks: A Century of Skiing in Utah*. Salt Lake City: Gibbs Smith.

74. Fey, M. 1983. *Slot Machines: An Illustrated History of America's Most Popular Coin-operated Gaming Device.* Las Vegas: Nevada Publishing.

75. Findlay, J. M. 1986. *People of Chance: Gambling in American Society from Jamestown to Las Vegas*. New York: Oxford University Press.

76. Fouberg, E. H. 1996. South Dakota Gaming: A Regional Analysis. *Great Plains Research* 6(2): 179–212.

77. Gargiulo V. 2005. *Palisade Amusement Park: A Century of Fond Memories*. Chicago: Arcadia Publishing.

78. Geist, C. D. 1994. Living History Villages as Popular Entertainments. *New England Journal of History* 51(2): 57–66.

79. Green, H. 1986. *Fit for America: Health, Fitness, Sport and American Society*. New York: Pantheon.

80. Grover, K., ed. 1989. *Fitness in American Culture: Images of Health, Sport and the Body 1830–1940.* Amherst: University of Massachusetts Press.

81. Guenther, K. 1998. Judge Roy's Playground: A History of Astroworld. *East Texas Historical Journal* 36(2): 58–67.

82. Harley, R. 2000. Roller Coaster Planet. *Convergence* 6(2): 77–95.

83. Hogan, D. G. 1998. Sea World: A New Breed of Amusement. *Reviews in American History* 26(4): 772–777.

84. Hurley, A. 2002. *Diners, Bowling Alleys and Trailer Parks: Chasing the American Dream in Postwar American Consumer Culture*. New York: Basic Books.

85. King, M. J. 1981. The New American Muse: Notes on the Amusement/Theme Park. *Journal of Popular Culture* 15(1): 56–62.

86. Kling, D. 2000. *The Rise of the Biggest Little City: An Encyclopedic History of Reno Gaming, 1931–1981*. Reno: University of Nevada Press.

87. LaPolla, F. 1988. The Psycho Geography of American Amusement Parks. *Reveue Francaise d'etudes Americaines* 8(36): 235–240.

88. Mason, W. D. 2000. *Indian Gaming: Tribal Sovereignty and American Politics*. Norman: University of Oklahoma Press.

89. Mayo, J. M. 1998. *The American Country Club: Its Origins and Development*. New Brunswick, NJ: Rutgers University Press.

90. Mills, S. F. 1992. The Contemporary Theme Park and Its Victorian Pedigree. *European Contributions to American Studies* 24: 78–96.

91. Mohun, A. P. 2001. Designed for Thrills and Safety: Amusement Parks and the Commodification of Risk, 1880–1929. *Journal of Design History* 14(4): 291–306.

92. Moss, R. J. 2001. *Golf and the American Country Club.* Urbana: University of Illinois Press.

93. Moss, R. 1993. Sport and Social Status: Golf and the Making of the Country Club in the United States, 1882–1920. *International Journal of the History of Sport* 10(1): 93–100.

94. Neil, J. M. 1981. The Roller Coaster: Architectural Symbol and Sign. *Journal of Popular Culture* 15(1): 108–115.

95. Nye, R. B. 1981. The American Amusement Park: Eight Ways of Looking at an Amusement Park. *Journal of Popular Culture* 15: 63.

96. Park, R. J. 1993. A Decade of the Body: Researching and Writing about the History of Health, Fitness, Exercise and Sport, 1983–1993. *Journal of Sport History* 21(1): 59–82.

97. Raento, P. 2000. Casino Gaming in the United States: Issues, Concerns, and Recent Research. *American Studies in Scandinavia* 32(2): 45–62.

98. Register, W. 2001. *The Kid of Coney Island: Fred Thompson and the Rise of American Amusements.* New York: Oxford University Press.

99. Roest, H., et al. 1997. Satisfaction with Amusement Parks. *Annals of Tourism Research* 24(4): 1001–1005.

100. Silliman, L. 1979. As Kind and Generous a Host as Ever Lived: Howard Eaton and the Birth of Western Dude Ranching. *American West* 16(4): 18–22.

101. Snow, R. E. 1976. Coney Island: A Case Study in Popular Culture and Technical Change. *Journal of Popular Culture* 9: 960.

102. Swarbrooke, J. 1995. *Development and Management of Visitor Attractions*. Burlington, MA: Butterworth-Heinemann.

103. Thomas, G. C. *Golf Architecture in America: Its Strategy and Construction*. Chelsea, MI: Sleeping Bear.

104. Weinstein, R. M. 1992. Disneyland and Coney Island: Reflections on the Evolution of the Modern Amusement Park. *Journal of Popular Culture* 26: 131.

READINGS IN HAZARDOUS WASTE MANAGEMENT AND POLLUTION PREVENTION

105. Balogh, J. C. and Walker, W. J. 1992. *Golf Course Management & Construction: Environmental Issues*. Boca Raton, FL: Lewis Publishers.

106. Dolgen, D., et al. 2003. Best Waste Management Programs for Marinas: A Case Study. *Journal of Coastal Conservation* 9(1): 57–63.

107. Impact Research Associates. 1995. *Clean Marinas: Clear Value Literature Review and Bibliography*. A Report Prepared for the U.S. EPA. Available online at www.epa.gov/owow/nps/marinas/appxa.html.

108. New Hampshire Department of Environmental Services. 2001. *Best Management Practices for New Hampshire Marinas* (NHDES-WD-01–12). Concord, NH: Pollution Prevention Program.

109. Rabasca, L. 1993. Disney World's Best Kept Secret Is Recycling. *Waste Age* 24(8): 28.

110. Scottish Environment Protection Agency. 2004. *Pollution Prevention Guidelines: Marinas and Craft* (PPG14). Stirling, UK: SEPA

READINGS IN HEALTH AND SAFETY

111. Funk, J. B. 1993. Reevaluating the Impact of Video Games. *Clinical Pediatrics* 32(2): 86–90.

112. Huang, P. P. 2003. Roller Coaster Headaches Revisited. *Surgical Neurology* 60(5): 398–401.

113. Jacobson, J. A., et al. 2005. Golf and Racquet Sports Injuries. *Seminars in Musculoskeletal Radiology* 9(4): 346–359.

114. Kim, D. H., et al. 2004. Shoulder Injuries in Golf. *American Journal of Sports Medicine* 32(5): 1324–1330.

115. Knopper, L. and Lean, D. R. 2004. Carcinogenic and Genotoxic Potential of Turf Pesticides Commonly Used on Golf Courses. *Journal of Toxicology & Environmental Health B. Critical Review* 7(4): 267–279.

116. Lester, D. and Jason, D. 1989. Suicides at the Casino. *Psychological Reports* 64(1): 337–338.

117. Pelletier, A. R. and Gilchirst, J. 2005. Roller Coaster Related Fatalities, U.S. 1994–2004. *Injury Prevention* 11(5): 309–312.

118. Pelton, H. K. 1999. Theme Park Noise Control. *Noise Control Engineering Journal* 47(5): 169–172.

119. Philips, J. H. 1995. Theme Park EMS. It's Not Mickey Mouse Operation. *JEMS* 20(7): 58–67.

120. Theriault, G. and Lachance, P. 1998. Golf Injuries. An Overview. *Sports Medicine* 26(1): 43–57.

Sector 72: Accommodation and Food Services

721: Accommodation
7211: Traveler Accommodation
7212: RV (Recreational Vehicle) Parks and Recreational Camps
7213: Rooming and Boarding Houses

READINGS IN HISTORY

1. Belasco, W. J. 1977. Americans on the Road: Auto Camping, Tourist Camps, Motels, 1910–1945. Dissertation, University of Michigan.
2. Bernstein, R. A. 1984. Boarding House Keepers and Brothel Keepers in New York City, 1880–1910. Dissertation, Rutgers University.
3. Bracken, C. E. 1997. Consuming Luxury: Hotels and the Rise of Middle Class Public Space, 1825–1860. Dissertation, George Washington University.
4. Braden, S. R. 2002. *The Architecture of Leisure: The Resort Hotels of Henry Fagler and Henry Plant.* Gainesville: University Press of Florida.
5. Cerullo, M. and Ewen, P. 1982. Having a Good Time: The American Family Goes Camping. *Radical America* 16(1–2): 13–44.
6. Colihan, J. 2001. Cradle of the Bed-and-Breakfast. *American Heritage* 52(3): 38–40.
7. Crain, J. 1977. *Historic Country Inns of California*. San Francisco: Chronicle Books.

8. Echeverria, J. 2000. Expansion and Eclipse of the Basque Boarding House in the American West. *Nevada Historical Society Quarterly* 43(2): 127–139.

9. Frisbie, L. K. 1983. Come In and Be Our Guest: Historic Hotels along the Suncoast. *Tampa Bay History* 5(2): 42–55.

10. Garwood, E. 1956. Early Texas Inns: A Study in Social Relationships. *Southwestern Historical Quarterly* 60(2): 219–244.

11. Getz, G. M. 1985. The Great Escape: Camping in 19th Century. *Pennsylvania Heritage* 11(3): 18–25.

12. Groth, P. 1994. *Living Downtown: The History of Residential Hotels in the United States.* Berkeley: University of California Press.

13. Groth, P. E. 1984. Forbidden Housing: The Evolution and Exclusion of Hotels, Boarding Houses, Rooming Houses, and Lodging Houses in American Cities, 1880–1930. Dissertation, University of California, Berkeley.

14. Hardwick, S. W. and Holtgrieve, D. G. 1991. Relics of a Time of Turbulence: Auto Camps by the River. *Californians* 9(3): 14–18.

15. Harmon, D. L. 2001. American Camp Culture: A History of Recreational Vehicle Development and Leisure Camping in the United States, 1890–1960. Dissertation, Iowa State University.

16. Harris R. 1992. The End Justified the Means: Boarding and Rooming in a City of Homes, 1890–1951. *Journal of Social History* 26(2): 331–358.

17. ———. 1994. The Flexible House: The Housing Backlog and the Persistence of Lodging, 1891–1951. *Social Science History* 18(1): 31–53.

18. Hart, J. F. 1984. Resort Areas in Wisconsin. *Geographical Review* 74(2): 192–217.

19. Hoffman, A. 1976. Mountain Resorts and Trail Camps in Southern California's Great Hiking Era, 1884–1938. *Southern California Quarterly* 58(3): 381–406.

20. Jakle, J. A., et al. 1996. *The Motel in America.* Baltimore, MD: Johns Hopkins University Press.

21. Kabat, S. G. 1995. Home Away from Home: The Architectural Geography of Hotels in the American Southwest, 1880–1920. Dissertation, University of New Mexico.

22. King, D. E. 1956. Early Hotel Entrepreneurs and Promoters, 1793–1860. *Explorations in Entrepreneurial History* 8(3): 148–160.

23. Lancaster, P. 1982. The Great American Hotel. *American Heritage* 33(4): 100–108.

24. Linden, B. M. G. 1981. Inns to Hotels in Cincinnati. *Cincinnati Historical Society Bulletin* 39(2): 127–152.

25. Lynn, S. D. 1999. *Windows on the Past: Historic Lodgings of New Mexico.* Albuquerque: University of New Mexico Press.

26. Mings, R. C. and McHugh, K. E. 1989. The RV Resort Landscape. *Journal of Cultural Geography* 10(1): 35–49.

27. Packard, J. C. 1968. San Diego's Early Hotels. *Southern California Quarterly* 59(3): 267–278.

28. Peel, M. 1986. On the Margins: Lodgers and Boarders in Boston, 1860–1900. *Journal of American History* 72(4): 813–834.

29. Reed, D. B. 2002. Life on Wheels: Camping in Pennsylvania. *Pennsylvania Heritage* 28(3): 14–21.

30. Roehl, W. S. and VanDoren, C. S. 1990. Locational Characteristics of American Resort Hotels. *Journal of Cultural Geography* 11(1): 71–83.

31. Rosenthal, L. S. 1967. Hotels: Pioneers in Progress. *American History Illustrated* 1(9): 42–53.

32. Scheer, V. 2000. The Farmhouse as Boarding House: Sharing Values and Sharing Space in the Catskills. *Voices* 26: 18–23.

33. Schrock, H. D. 2003. A Room for the Night: Evolution of Roadside Lodging in Wyoming. *Annals of Wyoming* 75(4): 31–39.

34. Sculle, K. A. 2000. Production of the Downtown Motel Chain in Memphis. *West Tennessee Historical Society Papers* 54: 94–108.

35. ——. 2002. Traveling in Style: The Park Plaza Motel Chain. *Journal of the West* 41(4): 63–70.

36. Spitulnik, K. M. 1972. The Inn Crowd: The American Inn, 1730–1830. *Pennsylvania Folklife* 22(2): 25–41.

37. Tolles, B. F. 1998. *The Grand Resort Hotels of the White Mountains: A Vanishing Architectural Legacy*. Boston: Godine.

38. Van Orman, R. A. 1968. Hotels of the Old West. *American History Illustrated* 3(6): 38–44.

39. Vieyra, D. I. 1995. The Architecture of America's Roadside Lodging from its Beginning to the Interstate Era. Dissertation, Case Western Reserve University.

40. Wharton, A. J. 2001. *Building the Cold War: Hilton International Hotels and Modern Architecture*. Chicago: University of Chicago Press.

41. Watkins, T. H. 1976. Home on the Road. *American West* 13(5): 36–48.

42. Wheeler, S. L. 1995. To Get an Honest Livin': Professional Boarding House Keepers in Hartford, 1890. *Connecticut History* 36(1): 48–65.

43. White, R. B. 1985. At Home on the Highway. *American Heritage* 37(1): 98–105.

44. ——. 2000. *Home on the Road: The Motor Home in America*. Washington DC: Smithsonian Institute.

45. Wriston, J. A. 1991. *Vermont Inns and Taverns: Pre-Revolution to 1925: An Illustrated and Annotated Checklist*. Rutland, VT: Academy Books.

46. Yoder, P. 1963. The American Inn, 1775–1850: Melting Pot or Stewing Kettle? *Indiana Magazine of History* 59(2): 135–151.

READINGS IN HAZARDOUS WASTE MANAGEMENT AND POLLUTION PREVENTION

47. Faulk, E. S. 2000. A Survey of Environmental Management by Hotels and Related Tourism Businesses. A Paper Presented at the OIKOS Ph.D. Summer Academy, University of St. Gallen.

48. International Hotels & Restaurant Association. 1996. *Environmental Good Practice in Hotels: Case Studies*. Paris: IHRA.

49. Kirk, D. 1997. *Environmental Management for Hotels*. Oxford: Butterworth-Heinemann.

50. New Hampshire Department of Environmental Services. 2001. *Staying Green: A Guide to Waste Management for the Lodging Industry in New Hampshire*. http://des.nh.gov/organization/divisions/waste/swrtas/documents/staying_green_appendix_d.pdf (accessed 7 November 2008).

51. Webster, K. 2000. *Environmental Management in the Hospitality Industry*. New York: Cassell; Guidotti F.P., 1960.

READINGS IN HEALTH AND SAFETY

52. Chyuan, J. Y., et al. 2004. Muskoskeletal Disorders in Hotel Restaurant Workers. *Occupational Medicine* 54(1): 55–57.

53. Enz, C. A. and Taylor, M. S. 2002. The Safety and Security of U.S. Hotels: A Post-September 11 Report. *Cornell Hotel & Restaurant Administration Quarterly* 43(5): 119–136.

54. Green, C. B., et al. 2004. Social Capital and the Value of Hunting Club Memberships. *Human Dimensions of Wildlife* 9(1): 57–68.

55. Kane, J. J. and Personick, M. E. 1993. Profiles in Safety and Health: Hotels and Motels. *Monthly Labor Review* 116(7): 36.

56. Louderback, J. 1996. Hotel Safety in the States. *Fire International* 153: 24–25.

57. Shaffer, H. J. and Hall, M. N. 2002. The Natural History of Gambling and Drinking Problems among Casino Employees. *Journal of Social Psychology* 142(4): 405–424.

58. Tener, S. and Ontko, J. 1976. *An Analysis of the Institutional and Commercial Housekeeper Occupation*. Columbus: Instructional Materials Laboratory, Trade and Industrial Education, Ohio State University.

59. Trout, D., et al. 1998. Exposure of Casino Employees to Environmental Tobacco Smoke. *Journal of Occupational & Environmental Medicine* 40(3): 270–276.

722: Food Services and Drinking Places
7221: Full-Service Restaurants
7222: Limited-Service Eating Places
7223: Special Food Services
7224: Drinking Places (Alcoholic Beverages)

READINGS IN HISTORY

60. Barbas, M. 2003. I'll Take Chop Suey: Restaurants as Agents of Culinary and Cultural Change. *Journal of Popular Culture* 36(4): 669–686.

61. Barry, P. 1927. Cafeteria. *American Speech* 3(1): 35–37.

62. Barton, R. J. 1992. Dining Railway Style. *Journal of the West* 31(1): 31–35.

63. Belasco, W. 2001. Roadside Dreams, Fast Food Nightmares. *Technology and Culture* 42(4): 767–770.

64. Bertrand, D. 1998. Mobile Canteen Industry in the United States. Dissertation, Concordia University, Montreal, Canada.

65. Bromell, N. 2000. The Automat: Preparing the Way for Fast Food. *New York History* 81(3): 300–312.

66. Everest, A. S. 1969. Early Roads and Taverns of the Champlain Valley. *Vermont History* 37(4): 247–255.

67. Garvin, J. L. 1995. Early White Mountain Taverns. *Historical New Hampshire* 50(1–2): 22–37.

68. Gerlach, J. D. 1991. The Prohibition Tavern: A Nebraska Study. *Journal of the West* 30(4): 67–75.

69. Gorelick, S. 1997. Big Mac Attacks: Lessons from the Burger Wars. *Ecologist* 27(5): 173–174.

70. Hathaway, J. 1986. A History of the American Drinking Place. *Landscape* 29(1): 1–9.

71. Hess, A. 1986. The Origin's of McDonalds' Golden Arches. *Journal of the American Society of Architectural Historians* 45(1): 60–67.

72. Hurley, A. 1997. From Hash House to Family Restaurant: The Transformation of the Diner and Post–World War II Consumer Culture. *The Journal of American History* 83(4): 1282–1308.

73. Jakle, J. A. and Sculle, K. A. 1999. *Fast Food: Roadside Restaurants in the Automobile Age.* Baltimore, MD: Johns Hopkins University Press.

74. Kaufmann, P. J. 1988. *Pizza Hut: Home Delivery.* Boston: Harvard Case Services.

75. Kindake, P. T. and Katovich, M. A. 1997. The Driver: Adaptations and Identities in the Urban World of Pizza Delivery Employees. *Journal of Contemporary Ethnography* 25(4): 421–448.

76. Langdon, P. 1986. *Orange Roofs, Golden Arches: The Architecture of American Chain Restaurants.* New York: Knopf.

77. Lohof, B. A. 1978. Hamburger Stand: Industrialization and the American Fast Food Phenomenon. *Industrial Archeology Review* 2(3): 265–276.

78. Luxenberg, S. 1985. *Roadside Empires: How the Chains Franchised America.* New York: Viking Penguin.

79. Manzo, J. T. 1990. From the Pushcart to Modular Restaurant: The Diner on the Landscape. *Journal of American Culture* 13(3): 13–21.

80. Mariani, J. 1991. *America Eats Out: An Illustrated History of Restaurants, Taverns, Coffee Shops, Speakeasies, and Other Establishments.* New York: William Morrow.

81. Parsa, H. G. and Kwansa, F. A. 2002. *Quick Service Restaurants, Franchising and Multi-Unit Chain Management.* New York: Haworth Hospitality Press.

82. Pillsbury, R. 1990. *Body Food, Soul Food: The American Restaurant Then and Now.* Boston: Unwin Hyman.

83. Rice, K. S. 1983. *Early American Tavern: For the Entertainment of Friends and Strangers*. Chicago: Regnery-Gateway.

84. Richards, C. E. 1963. City Taverns. *Human Organization* 22(4): 260–268.

85. Salinger, S. V. 2002. *Taverns and Drinking in Early America*. Baltimore, MD: Johns Hopkins University Press.

86. Shuldiner, A. T. 2001. Trapped behind the Automat: Technological Systems and the American Restaurant, 1902–1911. Dissertation, Cornell University, Ithaca.

87. Tangires, H. 1990. American Lunch Wagons. *The Journal of American Culture* 13(2): 91–108.

88. Whyte, W. F. 1949. The Social Structure of the Restaurant. *The American Journal of Sociology* 54(4): 302–310.

READINGS IN HAZARDOUS WASTE MANAGEMENT AND POLLUTION PREVENTION

89. Cummings, L. E. 1994. Solid Waste Minimization and the Foodservice Marketing Matrix. *Journal of Restaurant & Foodservice Marketing* 1: 57–73.

90. Full Circle Resources. 1995. *Restaurant Waste Reduction Manual*. Kansas City, MO: Bridging the Gap.

91. Georgia Pollution Prevention Assistance Division, Dept of Natural Resources. A fact sheet for food service waste reduction. http://www.p2ad.org/files_pdf/food_serv_wreduct.pdf (accessed 7 November 2008).

92. King County Solid Waste Division. n.d. *Recipes to Prevent Waste in the Restaurant.* http://www.p2pays.org/ref/20/19006.pdf (accessed 7 November 2008).

READINGS IN HEALTH AND SAFETY

93. Beckers, H. J. 1988. Microbiology and Food Hygiene in Mass Catering. *Catering Health* 1(1): 3–5.

94. Cash, A., et al. 2004. Smoking in Bars and Restaurants. *Journal of the Oklahoma State Medical Association* 97(8): 392–33.

95. CCOHS. *Food Service Workers Safety Guide*. http://www.escaladetraining.ca/PDF/food%20service%20sampleguide.pdf (accessed 7 November 2008).

96. Contato, R., et al. 2001. Nutritional Evaluation of Some Processed Catering Foods. *International Journal of Food Science & Nutrition* 52(1): 71–77.

97. Deeming, C. 2002. Hospital Catering. Hard to Swallow. *Health Service Journal* 112(5819): 28–29.

98. Eisner, M. D., et al. 1998. Bartender's Respiratory Health after Establishment of Smoke-free Bars and Taverns. *JAMA* 280(22): 1909–1914.

99. Gleeson, D. 2001. Health and Safety in the Catering Industry. *Occupational Medicine* 51: 385–391.

100. Grattan, M. 1998. All Vegetarian Cafeterias and Catering Pose Unique Challenges for Hospitals. *Health Foodservice* 8(3): 1–11.

101. Johnsson, K. O. and Berglund, M. 2003. Education of Key Personnel in Student Pubs Leads to a Decrease in Alcohol Consumption among the Patrons. *Addiction* 98(5): 627–633.

102. Jones, T. F., et al. 2004. Restaurant Inspection Scores and Food Borne Disease. *Emerging Infectious Disease* 10(4): 688–692.

103. Miquel, J. P. 1983. The Amalgamation of American and European Methods in the Catering Industry. *Journal of the Royal Society of Health* 103(2): 68–70.

104. Moy, G., et al. 1997. Improving the Safety of Street-vended Food. *World Health Statistics Quarterly* 50(1–2): 124–131.

105. Mubarak, S. J., et al. 1998. Ice-cream Truck Related Injuries to Children. *Journal of Pediatric Orthopedics* 18(1): 46–48.

106. Nichols, J. L. 1987. Floor Safety in Commercial Kitchens. *Journal of the Royal Society of Health* 107(4): 148–150.

107. Oliver, H., et al. 2005. Clustered Outbreak of Skin and Eye Complaints among Catering Staff. *Occupational medicine* 55(2): 149–153.

108. Peters, W. and Knighton, J. 1993. Burns Caused by Flambé Foods. *Journal of Burn Care Rehabilitation.* 14(3): 382–384.

109. Shepherd, J. P. and Warburton, A. 2000. Toughened Glassware and Injuries in Bars. *Injury Prevention* 6(3): 239–240.

110. Skroder, P. 1988. Hygiene—An Economic Factor in Catering. *Catering Health* 1(1): 7–10.

111. Wood, B. P. and Greig, D. E. 1997. Occupational Dermatitis—What's New? *Catering Industry Clinical Dermatology* 15(4): 567–571.

112. Zifferblatt, S. M., et al. 1980. A New Direction for Public Health Care: Changing Cafeteria Eating Habits. *Journal of the American Dietetic Association* 76(1): 15–20.

Sector 81: Other Services (Except Public Administration)

811: Repair and Maintenance
8111: Automotive Repair and Maintenance
8112: Electronic and Precision Equipment Repair and Maintenance
8113: Commercial and Industrial Machinery and Equipment (Except Automotive and Electronic) Repair and Maintenance
8114: Personal and Household Goods Repair and Maintenance

READINGS IN HISTORY

1. Aduddell, R. M. and Cain, L. P. 1990. Radio, TV, and Communication Equipment. In *Manufacturing: A Historiographical and Bibliographical Guide,* ed. D. O. Whitten, 320–328. Westport, CT: Greenwood Press.

2. Barnes, N. W. and Lyon, L. S. 1922. Marketing the Stephens Brake Shoe. *The University Journal of Business* 1(1): 95–113.

3. Borg, K. L. 2000. From the Village Blacksmith to Mr. Good Wrench: Creating Auto Mechanics in Technology's Middle Ground. Dissertation, University of Delaware.

4. Crandall, R. W. 1970. The Decline of the Franchised Dealer in the Automobile Repair Market. *Journal of Business* 43(1): 19–30.

5. Gerner, J. L. and Bryant, W. K. 1980. The Demand for Repair Service during Warranty. *The Journal of Business* 53(4): 397–414.

6. Goat, L. G. 1989. Housing the Horseless Carriage: America's Early Private Garages. *Perspectives in Vernacular Architecture* 3: 62–72.

7. Kuns, R. F. 1923. *Automotive Trade Training*. Milwaukee: Bruce Publishing.

8. Lenk, J. D. 1977. *Handbook of Basic Electronic Troubleshooting*. Englewood Cliffs, NJ: Prentice-Hall.

9. McIntrye, S. L. 1997. The Repair Man Will Gyp You: Mechanics, Managers and Customers in the Automobile Repair Industry, 1896–1940. Dissertation, University of Missouri, Columbia.

10. ——. 2000. The Failure of Fordism: Reform of the Automobile Repair Industry, 1913–1940. *Technology and Culture* 41(2): 269–299.

11. Nelson, B. J. 1998. The Nature and Implications of Technological Change and the Rise of a Service Economy: Observations from the Field of Automotive Repair. Dissertation, Cornell University.

12. Watkins, E. 1991. "For the Time Being, Forever": Social Position and the Art of Automobile Maintenance. *Boundary* 18(2): 150–165.

READINGS IN HAZARDOUS WASTE MANAGEMENT AND POLLUTION PREVENTION

13. Alabama Department of Environmental Management. 1990. *Automotive Refinishing Shops: Best Management Practices.* Montgomery, AL: Pollution Prevention Unit.

14. Arizona Department of Environmental Quality. 1998. *Pollution Prevention for Automotive Maintenance & Repair Shops.* Glendale, AZ: Pollution Prevention Unit.

15. Georgia Tech Research Institute. 1998. *Automotive Repair and the Environment. A Report Prepared for U.S. EPA, Region 4*. Atlanta: Georgia Tech Research Institute.

16. Kansas Department of Health & Environment. 1999. *Auto Body Shops: A Primer on Environmental Regulation & Pollution Prevention.* http://www.sbeap.org/publications/autobody.pdf (accessed 11 November 2008).

17. Kansas State University. n.d. *Pollution Prevention for the Automotive Maintenance & Repair Industry*. Manhattan, KS: Pollution Prevention Institute.

18. New Hampshire Department of Environmental Services. n.d. *Pit Stops Manual: Best Management Practices for Automobile Service Facilities.* Concord, NH: Pollution Prevention Program.

19. Rifkin, V. H. 2001. The Automotive Repair Industry: Pollution and Prevention. Dissertation, New York Institute of Technology, Old Westbury.

20. Scottish Environmental Protection Agency. 2004. *Pollution Prevention Guidelines: Garages and Vehicle Service Centers* (PPG 19). http://publications.environment-agency.gov.uk/pdf/PMHO06CATRO-e-e.pdf (accessed 11 November 2008).

21. U.S. Environmental Protection Agency. 1991. *Guides to Pollution Prevention: The Automotive Refinishing Industry* (EPA/625/7–91/016). Washington, DC: Office of Research & Development.

22. ——. 1997. *Automotive Refinishing Industry: Isocyanates Profile* (EPA/744/R-97/007). http://www.epa.gov/dfe/pubs/auto/profile/isocynates-revision4b1c.pdf (accessed 11 November 2008).

23. ——. 1998. *Auto Body Refinishing General Best Shop Practices* (EPA/744/R-98/008). Washington, DC: EPA.

24. ——. 1999. *RCRA in Focus: Vehicle Maintenance* (EPA/530/K-99/004). Washington, DC: Office of Solid Waste & Emergency Response.

25. ——. 2000. *Best Practices for Auto Refinishers when Spray Painting* (EPA/744/F-00/002). http://www.epa.gov/dfe/pubs/auto/bp_spray/bp_spray.pdf (accessed 19 November 2008).

26. ——. 2003. *Consolidated Screening Checklist for Automotive Repair Facilities Guidebook*. http://www.ccar-greenlink.org/vshops/FinalAutoGuide_092503.pdf (accessed 19 November 2008).

27. Washington State Department of Ecology. 1999a. *A Guide for Automotive Machine Shops* (92–BR-11). http://www.ecy.wa.gov/pubs/92br11.pdf (accessed 19 November 2008).

28. Washington State Department of Ecology. 1999b. *A Guide for Radiator Shops* (92–BR-9). http://www.ecy.wa.gov/pubs/92br9.pdf (accessed 19 November 2008).

29. Washington State Department of Ecology. 1999c. *A Guide for Transmission Shops* (92–BR-10). http://www.ecy.wa.gov/pubs/92br10.pdf (accessed 19 November 2008).

30. Washington State Department of Ecology. 2003. *A Guide for Auto Body Shops*. http://www.ecy.wa.gov/pubs/92br16.pdf (accessed 19 November 2008).

READINGS IN HEALTH AND SAFETY

31. Antonini, J. M. 2003. Health Effects of Welding. *Critical Reviews in Toxicology* 33: 61–103.

32. Barbee, J. Y. and Prince, T. S. 1999. Acute Respiratory Distress Syndrome in a Welder Exposed to Metal Fumes. *Southern Medical Journal* 92(5): 510–512.

33. Daniell, W., et al. 1993. Neuropsychological Performance and Solvent Exposure among Car Body Repair Shop Workers. *British Journal of Industrial Medicine* 50: 368–377.

34. Friede, E. 1961. Symptomatic Pulmonary Disease in Arc Welders. *Annals of Internal Medicine* 54(1): 121–127.

35. Harbine, K. L. 2003. Health Hazards in Automobile Repair Shops. *Clinical Occupational and Environmental Medicine* 3: 45–59.

36. Hewitt, P. J. 2001. Strategies for Risk Assessment and Control in Welding: Challenges for Developing Countries. *Annals of Occupational Hygiene* 45: 295–298.

37. Jayjock, M. A. and Levin, L. 1984. Health Hazards in a Small Auto Body Repair Shop. *Annals of Occupational Hygiene* 28(1): 19–29.

38. Korczynski, R. E. 2000. Occupational Health Concerns in the Welding Industry. *Applied Occupational Environmental Hygiene* 15(12): 936–945.

39. Matczak, W. and Gromiec, J. 2002. Evaluation of Occupational Exposure to Toxic Metals Released in the Process of Aluminum Welding. *Applied Occupational Environmental Hygiene* 17(4): 296–303.

40. Moen, B. E. and Hollund, B. E. 2000. Exposure to Organic Solvents among Car Painters in Bergen, Norway. *Annals of Occupational Hygiene* 44(3): 185–189.

41. Niemeal, R., et al. 1987. Ventilation and Organic Solvent Exposure during Car Washing. *Scandinavian Journal of Work & Environmental Health* 13(5): 424–430.

42. NIOSH. 1975. *Asbestos Exposure During Servicing of Motor Vehicle Brake and Clutch Assemblies* (CIB No. 5). http://www.cdc.gov/niosh/78127_5.html (accessed 19 November 2008).

43. NIOSH. 1977. *Safety and Health in Auto Body Repair Shops* (NIOSH Pub. No. 77–2290). Cincinnati: NIOSH.

44. Oxhoj H., et al. 1979. Effects of Electric Arc Welding on Ventilatory Lung Function. *Archives of Environmental Health* 34: 211–217.

45. Parker, D. L., et al. 1991. A Cross-sectional Study of Pulmonary Function in Auto Body Repair Workers. *American Journal of Public Health* 81: 768–771.

46. Racette, B. A., et al. 2001. Welding-Related Parkinsonism. *Neurology* 56: 8–13.

47. Rose, D. N., et al. 1982. Screening for Lead Toxicity among Auto-body Repair Workers. *American Journal of Industrial Medicine* 3: 405–412.

48. Suruda, A., et al. 1991. Injuries from Tire and Wheel Explosions during Servicing. *Annals of Emergency Medicine* 20(8): 848–851.

49. Treble, R. G., et al. 1998. Elevated Blood Levels from Exposure via a Radiator Workshop. *Environmental Research* 77(1): 62–65.

50. U.S. Environmental Protection Agency. 2002. *Respiratory Protection Program for Auto Refinish Shops* (EPA/744/F-02/010). http://www.epa.gov/dfe/pubs/auto/respirator/rpp.pdf (accessed 19 November 2008).

51. Williams, N. R., et al. 1999. Biological Monitoring to Assess Exposure from Use of Isocyanates in Motor Vehicle Repair. *Occupational and Environmental Medicine* 56: 598–601.

52. Wong, O. 2001. Malignant Mesothelioma and Asbestos Exposure among Auto Mechanics: Appraisal of Scientific Evidence. *Regulatory Toxicology & Pharmacology* 34(2): 170–177.

812: Personal and Laundry Services
8121: Personal Care Services
8122: Death Care Services
8123: Dry Cleaning and Laundry Services
8129: Other Personal Services

READINGS IN HISTORY

53. Armour, R. A. and Williams, J. C. 1981. Image Making and Advertising in the Funeral Industry. *Journal of Popular Culture* 14(4): 701–710.

54. Begley, C. E. 1991. Government Should Strengthen Regulation in the Weight Loss Industry. *Journal of the American Dietetic Association* 91(10: 1255–1257.

55. Blackwell, R. D. 1967. Price Levels in the Funeral Industry. *Quarterly Review of Economics & Business* 7(4): 75–84.

56. Brand, H. and Ahmed, Z. Z. 1986. Beauty and Barber Shops: The Trend of Labor Productivity. *Monthly Labor Review* 109(3): 21–26.

57. Bristol, D. W. 2003. From Outposts to Enclaves: A Social History of Black Barbers, 1750–1915. Dissertation, University of Maryland.

58. Church, R. A. and Smith, B. M. D. 1966. Competition and Monopoly in the Coffin Furniture Industry, 1870–1915. *The Economic History Review* 19(3): 621–641.

59. Darnall, M. J. 1983. The American Cemetery as Picturesque Landscape. *Winterthur Portfolio* 18(4): 249–269.

60. Davis, D. J. and Mates, L. H. 2005. *Encyclopedia of Cremation*. Burlington, VT: Ashgate.

61. deHaas, E. 1946. Concepts of the Nature of Bail in English and American Criminal Law. *The University of Toronto Law Journal* 6(2): 385–400.

62. DeMello, M. 2000. *Bodies of Inscription: A Cultural History of the Modern Tattoo Community*. Durham, NC: Duke University Press.

63. Dukert, A. R. 1958. "Cleaners" vs. "Cleansers." *American Speech* 33(2): 158–159.

64. Eakin, J. S. 2005. *Salons and Spas: The Architecture of Beauty*. Beverly, MA: Rockport Publishers.

65. Embree, J. L. 1995. Wash Day Woes of the Textile Conservator: Laundry Methods of the Turn of the Century. *Ars Textrina* 23: 73–95.

66. Erickson, M. 2003. *Start Your Own Coin Operated Laundry*. Irvine, CA: Entrepreneur Press.

67. Flood, K. P. 2001. Contemplating Corpses: The Dead Body in American Culture, 1870–1920. Dissertation, Harvard University.

68. Foster, G. S. and Hummel, R. L. 1995. The Adkins-Woodson Cemetery: A Sociological Examination of Cemeteries as Communities. *Markers* 12: 92–117.

69. Fraser L. 1997. *Losing It: America's Obsession with Weight and the Industry That Feeds on It.* New York: Dutton.

70. Froeb, L., et al. 2003. Bertrand Competition with Capacity Constraints: Mergers among Parking Lots. *Journal of Econometrics* 113(1): 46–97.

71. Furman, F. K. 1997. *Facing the Mirror: Older Women and Beauty Shop Culture*. New York: Routledge.

72. Gimlin, D. 1996. Pamela's Place: Power and Negotiation in the Hair Salon. *Gender and Society* 10(5): 505–526.

73. Hijiya, J. A. 1982. American Gravestones and Attitudes toward Death: A Brief History. *Proceedings of the American Philosophical Society* 127(5): 339–363.

74. Jackson, K. T. and Vergara, C. J. 1989. *Silent Cities: The Evolution of the American Cemetery*. New York: Princeton Architectural Press.

75. James, R. M., et al. 1994. Competition and Coexistence in the Laundry: A View of the Comstock. *The Western Historical Quarterly* 25(2): 164–184.

76. Jett, S. C. 1996. Modern Navajo Cemeteries. *Material Culture* 28(2): 1–23.

77. Kopp, S. W. and Kemp, E. 2007. The Death Care Industry: A Review of Regulatory and Consumer Issues. *The Journal of Consumer Affairs* 41(1): 150–173.

78. Laderman, G. 2003. *Rest in Peace: A Cultural History of Death and the Funeral Home in 20th Century America.* New York: Oxford University Press.

79. Loose, J. W. W. 1985. The Lancaster Crematorium: First in the USA. *Journal of the Lancaster County Historical Society* 89(2): 46–52.

80. Magnuson, J. 1960. Parking Facilities: Some Legal and Financial Considerations. *Virginia Law Review* 46(3): 595–612.

81. Mayer, R. G. 2000. *Embalming: History, Theory & Practice.* New York: McGraw-Hill.

82. Meyer, R. E., ed. 1992. *Cemeteries and Grave Markers: Voices of American Culture.* Logan: Utah State University Press.

83. Mohun, A. P. 1992. Women, Work and Technology: The Steam Laundry Industry in the United States and Great Britain, 1880–1920. Dissertation, Case Western Reserve University.

84. Montgomery, S. W. 1999. Making Room for the Automobile: The Parking Garage in the City Center, 1920–1970. Dissertation, George Washington University.

85. Ong, P. 1983. Chinese Laundries as an Urban Occupation in 19th Century California. *Annals of the Chinese Historical Society of the Pacific Northwest* 1983: 68–85.

86. Parris, R. L. 1967. Distribution of Barber Shops in Washington, D.C.: An Example of Changing Urban Land Use. Dissertation, Catholic University of America.

87. Peiss, K. 2000. On Beauty . . . and the History of Business. *Enterprise and Society* 1(3): 485–506.

88. Pine, V. R. and Phillips, D. L. 1970. The Cost of Dying: A Sociological Analysis of Funeral Expenditures. *Social Problems* 17(3): 405–417.

89. Preston, C. A. 1989. Bail Bonds in the Criminal Justice Process: The Myth and the Reality. Dissertation, Baylor University, Waco, Texas.

90. Prothero, S. R. 2001. *Purified by Fire: A History of Cremation in America.* Berkeley: University of California Press.

91. Randall, H. W. 1997. Sun Tanning: Differences in Perceptions Throughout History. *Mayo Clinic Proceedings* 72(5): 461–466.

92. Richardson, D. J. and Markley, R. 1975. The Jargon of Barbershop. *American Speech* 50(3/4): 293–304.

93. Simpson, H. S. 1927. Downtown Storage Garages. *Annals of the American Academy of Political & Social Science* 133: 82–89.

94. Sloan, D. C. 1991. *The Last Great Necessity: Cemeteries in American History*. Baltimore, MD: Johns Hopkins University Press.

95. Stewart, J. A. 1900. The Model Public Bath at Brookline. *The American Journal of Sociology* 5(4): 470–474.

96. Thompson, A. L. 1998. Unveiled: The Emotion Work of Wedding Coordinators in the American Wedding Industry. Dissertation, Brandeis University.

97. Thompson, J. A. 1958. *A Study of the Philco-Bendix Commercial Coin Operated Laundry Program*. Salt Lake City: College of Business, University of Utah.

98. Thompson, S., et al. 1993. Historic Cemeteries in the Southern United States: A Preliminary Bibliography. *Southern Quarterly* 31(2): 133–146.

99. Tsai, A. G. and Walden, T. A. 2005. Systematic Review: An Evaluation of Major Commercial Weight Loss Programs in the United States. *ACP Journal Club* 143(1): 15.

100. Wilhelm, A. E. 1988. Pretty Is as Pretty Does: The Rhetoric of Beauty Salon Names. *Names* 36(1–2): 61–68.

101. Wilson, C. R. 1983. The Southern Funeral Director: Managing Death in the New South. *Georgia Historical Quarterly* 67(1): 49–69.

102. Wilson, W. A. 1988. Freeways, Parking Lots and Ice Cream Stands: The Three Nephrites in Contemporary Society. *Dialogue* 21(3): 13–26.

103. Wueschner, S. A. 2001. One Man's Demise Is Another Man's Gain: The Growth of the Funeral Industry on the Iowa Frontier. *Essays in Economic & Business History* 19: 245–251.

READINGS IN HAZARDOUS WASTE MANAGEMENT AND POLLUTION PREVENTION

104. Arnold, P. R. 1971. The History and Analysis of the Marketing of Perchloroethylene in the United States. Dissertation, McNeese State University.

105. Monson, B. R. 1983. A Study of Dry Cleaners as Generators of Small Quantities of Hazardous Waste. Dissertation, University of Cincinnati.

106. U.S. Environmental Protection Agency. 1990. *Dry Cleaning and Laundry Plants* (RCRA Information Sheet, EPA/530/SW-90/027b). Washington, DC: EPA.

107. ——. 1992. *Proceedings of the International Roundtable on Pollution Prevention & Control in the Dry Cleaning Industry* (EPA/774/R-92/002).Washington, DC: EPA.

108. ——. 1995. *Profile of the Dry Cleaning Industry* (EPA/310/R-95/001). Washington, DC: EPA.

109. ——. 1996a. *Plain English Guide for Perc Dry Cleaners* (EPA/305/B-96/002). Washington, DC: EPA.

110. ——. 1996b. *Proceedings: Apparel Care and the Environment* (EPA/744/R-96/002). Washington, DC: EPA.

111. ——. 1998. *Cleaner Technologies Substitutes Assessment for Professional Fabricare Process* (EPA/744/B-98/001). Washington, DC: EPA.

112. ——. 1999a. *RCRA in Focus: Dry Cleaning* (EPA/530/K-99/005). Washington, DC: EPA, Office of Solid Waste & Emergency Response.

113. ——. 1999b. *RCRA in Focus: Photo Processing* (EPA/530/K-99/002). Washington, DC: EPA, Office of Solid Waste & Emergency Response.

114. ——. 2001. *Garment and Textile Care Resource Guide* (EPA/744/K-98/005). Washington, DC: EPA.

READINGS IN HEALTH AND SAFETY

115. Blair, A., et al. 1979. Causes of Death among Laundry and Dry Cleaning Workers. *American Journal of Public Health* 69: 508–511.

116. ——. 1990. Cancer and Other Causes of Death among a Cohort of Dry Cleaners. *British Journal of Industrial Medicine* 47: 162–168.

117. Braithwaite, R. L., et al. 1999. Risk Associated with Tattooing and Body Piercing. *Journal of Public Health Policy* 20(4): 459–470.

118. Brown, D., et al. 1987. Retrospective Cohort Study of Dry Cleaner Workers Using Perchloroethylene. *Journal of Occupational Medicine* 29: 535–541.

119. Duh, R. W. and Asal, N. R. 1984. Mortality among Laundry and Dry Cleaning Workers in Oklahoma. *American Journal of Public Health* 74: 1278–1280.

120. Gulyas, H. and Hemmerling, L. 1990. Tetrachloroethene Air Pollution Originating from Coin-operated Dry Cleaning Establishments. *Environmental Research* 53(1): 90–99.

121. IARC. 1993. *IARC Monographs on the Evaluation of Carcinogenic Risks to Humans: Occupational Exposures of Hairdressers and Barbers and Personal Use of Hair Colorants; Some Hair Dyes, Cosmetic Colorants, Industrial Dyestuffs and Aromatic Amines* (Monograph No. 57). Lyon, France: IARC.

122. ——. 1995. *IARC Monographs on the Evaluation of Carcinogenic Risks to Humans: Dry Cleaning, Some Chlorinated Solvents and Other Industrial Chemicals* (Monograph No. 63). Lyon, France: IARC.

123. Katz, R. M. and Jowett, D. 1981. Female Laundry and Dry Cleaning Workers in Wisconsin: A Mortality Analysis. *American Journal of Public Health* 71: 305–307.

124. Levine, J., et al. 2005. The Indoor UV Tanning Industry: A Review of Skin Cancer Risk, Health Benefit Claims and Regulation. *Journal of the American Academy of Dermatology* 53(6): 1038–1044.

125. NIOSH. 1976. *Criteria for a Recommended Standard: Occupational Exposure to Trichloroethylene* (HEW Publication No. 76–185). Washington, DC: NIOSH.

126. NIOSH. 1997. *Control of Health & Safety Hazards in Commercial Drycleaners: Chemical Exposures, Fire Hazards, and Ergonomic Risk Factors* (DHHS Publication No. 97–150). http://www.cdc.gov/niosh/97-150.html (accessed 19 November 2008).

127. U.S. EPA. 1999. *Key Characteristics of Laundry Detergent Ingredients* (EPA/744/F-99/008). http://www.epa.gov/dfe/pubs/laundry/techfact/keychar.htm (accessed 19 November 2008).

128. WorkCover NSW. 2003. *Health and Safety Guidelines for Hairdressers.* Gosford, New South Wales: WorkCover NSW.

813: Religious, Grantmaking, Civic, Professional, and Similar Organizations

READINGS IN HISTORY

129. Abdul-Wajid Bagby, I., et al. 2001. *The Mosque in America: A National Portrait*. Washington, DC: Council on American-Islamic Relations.

130. Aksartova, S. 2003. In Search of Legitimacy: Peace Grant Making of U.S. Philanthropic Foundations: 1988–1996. *Nonprofit and Voluntary Sector Quarterly* 32(1): 25–46.

131. Aldrich, J. H. 1995. *Why Parties: The Origin and Transformation of Political Parties in America.* Chicago: University of Chicago Press.

132. Ashenfelter, O. and Pencavel, J. H. 1969. American Trade Union Growth: 1900–1960. *The Quarterly Journal of Economics* 83(3): 434–448.

133. Baker, H. A. 1911. Charitable Trusts in Iowa. Dissertation, Yale University, School of Law.

134. Baker, S. B. 1984. The Cleveland Chamber of Commerce in the Great Depression. Dissertation, Case Western Reserve University.

135. Becker, W. H. 1971. American Wholesale Hardware Trade Associations, 1870–1900. *The Business History Review* 45(2): 179–200.

136. Bloomfield, W. M. 2002. Grantmaking Foundations in America: Analyzing the Process and Practice of Philanthropic Decision Making. Dissertation, Brandeis University.

137. Bucher, R. and Stelling, J. 1969. Characteristics of Professional Organizations. *Journal of Health & Social Behavior* 10(1): 3–15.

138. Burleson, F. E. 1926. Community Chests in Small Cities. *Social Forces* 5(1): 85–90.

139. Castle, A. L. 2004. *A Century of Philanthropy: A History of the Samuel N. and Mary Castle Foundation*. Honolulu: Hawaiian Historical Society.

140. Charles, J. A. 1993. *Service Clubs in American Society: Rotary, Kiwanis and Lions*. Urbana: University of Illinois Press.

141. Coats, A. W. 1964. The American Economic Association, 1904–1929. *The American Economic Review* 54(4): 261–285.

142. Coon, H. 1990. *Money to Burn: Great American Foundations and Their Money*. New Brunswick, NJ: Transaction Publishers.

143. Curtie, M. 1957. The History of American Philanthropy as a Field of Research. *The American Historical Review* 62(2): 352–363.

144. Dimit. Factors Associated with Success and Failure of Pennsylvania Granges. Dissertation, Pennsylvania State University.

145. Dowie, M. 2001. *American Foundations: An Investigative History*. Cambridge, MA: MIT Press.

146. Elazar, D. J. 1984. The Development of the American Synagogue. *Modern Judaism* 4(3): 255–273.

147. Endress, R. M. 1974. The Enduring Vision: Stability Change in an American Benedictine Monastery. Dissertation, St. Martins College.

148. Estill, F. S. 1921. Organizing the Chamber of Commerce for Maximum Efficiency. Dissertation, University of Texas.

149. Ferguson, J. S. 1942. The Grange and Farmer Education in Mississippi. *The Journal of Southern History* 8(4): 497–512.

150. Finke, R. and Stark, R. 1992. *The Churching of America, 1776–1990: Winners and Losers in Our Religious Economy*. New Brunswick, NJ: Rutgers University Press.

151. Frank, R. G. and Salkever, D. S. 1994. Nonprofit Organization in the Health Sector. *The Journal of Economic Perspectives* 8(4): 129–144.

152. Freeman, M. 1994. The Philosophical Foundations of Human Rights. *Human Rights Quarterly* 16(3): 491–514.

153. Gauldfeldt, F. I. 1976. Interorganizational Relations of the United States Voluntary Health Organizations in Their Host Countries. Thesis. George Washington University.

154. Goode, W. J. 1957. Community within a Community: The Professions. *American Sociological Review* 22(2): 194–200.

155. Halliday, T. C., et al. 1987. Minimalist Organizations: Vital Events in State Bar Associations, 1870–1930. *American Sociological Review* 52(4): 456–471.

156. Hannan, M. T. and Freeman, J. 1988. The Ecology of Organizational Mortality: American Labor Unions, 1836–1985. *The American Journal of Sociology* 94(1): 25–52.

157. Hoxie, R. F. 1914. Trade Unionism in the United States. *The Journal of Political Economy* 22(3): 201–217.

158. Hudson, W. S. 1953. *The Great Tradition of the American Churches*. New York: Harper.

159. Kirby, A. J. 1988. Trade Associations as Information Exchange Mechanisms. *The RAND Journal of Economics* 19(1): 138–146.

160. Knoke, D. 1986. Associations and Interest Groups. *Annual Review of Sociology* 12: 1–21.

161. Levesque, S. M. 2000. From Yellowstone to Yukon: Combining Science and Advocacy to Shape Public Opinion. Dissertation, University of California, Irvine.

162. MacLeod, D. I. 1982. Act Your Age: Boyhood, Adolescence, and the Rise of the Boy Scouts of America. *Journal of Social History* 16(2): 3–20.

163. Marrett, C. G. 1980. Influences on the Rise of New Organizations: The Formation of Women's Medical Societies. *Administrative Science Quarterly* 25(2): 185–199.

164. McBride, H. W. 2000. Fraternal Regalia in America, 1865–1918: Dressing the Lodges; Clothing the Brotherhood. Dissertation, Ohio State University.

165. McCarthy, J. D. and Wolfson, M. 1996. Resource Mobilization by Local Social Movement Organizations. *American Sociological Review* 61(6): 1070–1088.

166. Mechanic, D. 1996. Changing Medical Organization and the Erosion of Trust. *The Milbank Quarterly* 74(2): 171–189.

167. Mendenhall, P. and Alexander, C. 1940. Guide to the Literature on the Boy Scouts of America. *The School Review* 48(5): 363–367.

168. Mogyorody, V. A. 1993. The Spectrum of Environmental Advocacy. Dissertation, Rensselaer Polytechnic Institute.

169. Oetgen, J. 1999. *Mission to America: A History of Saint Vincent Arch Abbey, the First Benedictine Monastery in the United States.* Catholic University of America Press.

170. Olmstead, C. E. 1960. *History of Religion in the United States*. Englewood Cliffs, NJ: Prentice-Hall.

171. Palmer, G. L. 1927. The Community Chest in Virginia Cities. *Social Forces* 6(2): 229–236.

172. Schmakel, E. C. 1968. *The Organization and Operation of Alumni Associations*. Toledo: University of Toledo.

173. Seville, M. A. 1987. Accounting and Information about Voluntary Health and Welfare Organizations. *Financial Accountability & Management* 3(3–4): 367–380.

174. Skocpol, T. and Oser, J. L. 2004. *Organization Despite Adversity: The Origins and Development of African American Fraternal Organizations.* Durham, NC: Duke University Press.

175. Smith, D. H. 1997. The International History of Grassroots Associations. *International Journal of Comparative Sociology* 38(3–4): 189–216.

176. Staber, U. H. 1982. *The Organizational Properties of Trade Associations*. Ithaca, NY: Cornell University Press.

177. Street, E. 1927. Professional Guidance for the Small Town Community Chest. *Social Forces* 5(4): 639–640.

178. Taft, P. 1964. *Organized Labor in American History*. New York: Harper & Row.

179. Tomlins, C. L. 1985. *The State and the Unions: Labor Relations, Law, and the Organized Labor Movement in America, 1880–1960*. New York: Cambridge University Press.

180. Walker, J. L. 1991. *Mobilizing Interest Groups in America: Patrons, Professions and Social Movements*. Ann Arbor: University of Michigan Press.

181. Walker, S. 1990. *In Defense of American Liberties: A History of the ACLU*. New York: Oxford University Press.

182. Werthheimer, J. 2003. *The American Synagogue: A Sanctuary Transformed*. New York: Cambridge University Press.

183. Willoughby, W. W. 1904. The American Political Science Association. *Political Science Quarterly* 19(1): 107–111.

184. Wilson, J. Q. 1974. *Political Organizations*. New York: Basic Books.

185. Wilson, L. G. 1926. Alumni Associations as Related to College Development. Dissertation, University of Kentucky, Lexington.

186. Wilt, M. 1995. A Study of the Practices of Successful Alumni Associations. Dissertation, Rowan College of New Jersey.

187. Wright, J. R. 1985. PACs, Contributions, and Roll Calls: An Organizational Perspective. *The American Political Science Review* 79(2): 400–414.

READINGS IN HEALTH AND SAFETY

188. Copping, A. G. 2002. The Development of a Fire Safety Evaluation Procedure for the Property Protection of Parish Churches. *Fire Technology* 38(4): 319–334.

189. Loving, A. L. and Wolf, L. 1997. The Effects of Receiving Holy Communion on Health. *Journal of Environmental Health* 60(July–August 1997): 6–10.

190. Sorace, S. and Terenzi, G. 2002. Seismic Evaluation and Retrofit of Historical Churches. *Structural Engineering International* 12(4): 283–288.

191. White H. H. 1968. Pack Palsy: A Neurological Complication of Scouting. *Pediatrics* 41(5): 1001–1003.

814: Private Households

READINGS IN HISTORY

192. Benn, D. W., et al. 1999. Prehistoric Households in Iowa. *Wisconsin Archeologist* 80(2): 161–210.

193. Bercaw, N. D. 2003. *Gendered Freedoms: Race, Rights and the Politics of Household in the Delta, 1861–1875*. Gainesville: University Press of Florida.

194. Birbari, E. L. 1951. Uniforms and Duties of Domestic Servants in America from 1890 to 1910. Dissertation, University of Texas.

195. Bowden, S. and Offer, A. 1994. Household Appliances and the Use of Time: The United States and Britain since the 1920s. *Economic History Review* 47(4): 725–748.

196. Busch, J. 1983. Cooking Competition: Technology on the Domestic Market in the 1930s. *Technology and Culture* 24(2): 222–245.

197. Fawver, K. M. 2002. Lords of Creation: Household and Family Structure in the 18th Century Chesapeake. Dissertation, University of California, Riverside.

198. Flannery, K. V. 2002. The Origins of the Village Revisited: From Nuclear to Extended Households. *American Antiquity* 67(3): 417–433.

199. Garcia, A. M. 1994. Maids No More: The Transformation of Domestic Work. *Frontiers: A Journal of Women Studies* 14(3): 171–180.

200. Haynes, E. R. 1923. Negroes in Domestic Service in the United States: Introduction. *Journal of Negro History* 8(4): 384–442.

201. Hoganson, K. 2002. Cosmopolitan Domesticity: Importing the American Dream, 1865–1920. *American Historical Review* 107(1): 55–83.

202. Kearny, P. J. 1993. The Nanny in the United States: A Search for Professional Recognition. Dissertation, University of Massachusetts, Amherst.

203. Kierner, C. A. 1998. *Beyond the Household: Women's Place in the Early South, 1700–1835*. Ithaca, NY: Cornell University Press.

204. Mokyr, J. 2000. Why "More Work for Mother?" Knowledge and Household Behavior, 1870–1945. *Journal of Economic History* 60(1): 1–41.

205. Ogle, M. 1993. Domestic Reform and American Household Plumbing, 1840–1870. *Winterthur Portfolio* 28(1): 33–58.

206. ——. 1996. *All the Modern Conveniences: American Household Plumbing, 1840–1890*. Baltimore, MD: Johns Hopkins University Press.

207. Parkhurst, J. W. 1938. The Role of the Black Mammy in the Plantation Household. *The Journal of Negro History* 23(3): 349–369.

208. Ruggles, S. and Browser, S. 2003. Measurement of Household and Family Composition in the United States, 1850–2000. *Population & Development Review* 29(1): 73–101.

209. Rutherford, J. W. 2003. *Selling Mrs. Consumer: Christine Frederick and the Rise of Household Efficiency*. Athens: University of Georgia Press.

210. Salmon, L. M. 1892. A Statistical Inquiry Concerning Domestic Service. *Publications of the ASA* 3(18/19): 89–118.

READINGS IN HAZARDOUS WASTE MANAGEMENT AND POLLUTION PREVENTION

211. Bolaane, B. 2004. Sampling Household Waste at Source: Lessons Learned at Gaborone. *Waste Management & Research* 22(3): 142–148.

212. Boldero, J. 1995. The Prediction of Household Recycling of Newspapers. *Journal of Applied Psychology* 25(5): 440–462.

213. Choe, C. and Fraser, I. 1998. The Economics of Household Waste Management: A Review. *Australian Journal of Agricultural and Resource Economics* 42(3): 269–302.

214. Emery, A. D., et al. 2003. An In-depth Study of the Effects of Socio-economic Conditions on Household Waste Recycling Practices. *Waste Management Research* 21(3): 180–190.

215. Jones, A., et al. 1999. Factors Affecting the Rate of Generation of Household Glass Waste. *Environmental & Waste Management* 2(2): 107–112.

216. Kettler, L. E., et al. 1997. Surveying Household Hazardous Waste Generation and Collection: Trends in Arizona. *Journal of Environmental Health* 59.

217. Ligman, K., et al. 1974. Household Wastewater Characterization. *Journal of Environmental Engineering* 100(1): 201–213.

218. Rufford, N. N. 1984. The Analysis and Prediction of the Quantity and Composition of Household Refuse. Dissertation, University of Aston, UK.

219. Slack, R. J., et al. 2004. Hazardous Components of Household Waste. *Critical Reviews in Environmental Science & Technology* 34(2004): 419–445.

220. Tucker, P. and Smith, D. 1999. Simulating Household Waste Management Behaviour. *Journal of Artificial Societies and Social Simulation* 2(3).

221. U.S. Environmental Protection Agency. 1993. *Household Hazardous Waste Management: A Manual for One Day Community Collection Programs* (EPA/530/R-92/026). Washington, DC: Office of Solid Waste & Emergency Response.

222. Zeiss, C. 1991. Predicting Household Hazardous Waste Generation Rates. *Journal of Environmental Systems* 21(3): 237–256.

READINGS IN HEALTH AND SAFETY

223. Anonymous. 2005. Household Safety: A Rough and Ready Guide. *Mayo Clinic Women's Healthsource* 9(3): 4–5.

224. Collia, D. V., et al. 2003. The 2001 National Household Travel Survey: A Look into the Travel Patterns of Older Americans. *Journal of Safety Research* 34: 461–470.

225. Cook, J. T. 2002. Clinical Implications of Household Food Security: Definitions, Monitoring, and Policy. *Nutrition & Clinical Care* 5(4): 152–167.

226. Dershewitz, R. A. 1979. Will Mothers Use Free Household Safety Devices? *American Journal of Disease in Children 1*33(1): 61–64.

227. Dershewitz, R. A. and Christophersen, E. R. 1984. Childhood Household Safety. An Overview. *American Journal of Diseases in Children* 138(1): 85–88.

228. Grey, C. N. B., et al. 2005. The Use and Disposal of Household Pesticides. *Environmental Research* 97(1): 109–15.

229. Heinen, M., et al. 2004. Injury Questions on Household Surveys from Around the World. *Injury Prevention* 10: 327–329.

230. Kassab, C., et al. 1995. The Changing Impact of Industry, Household Structure, and Residence on Household Well-being. *Rural Sociology* 60(1): 67–90.

231. Leonard, S. D. and Wogalter, M. S. 2000. What You Don't Know Can Hurt You: Household Products and Events. *Accident Analysis & Prevention* 32(3): 383–388.

232. Oh, H. J. 1998. The Effects of Household and Neighborhood Poverty on the Individual's Risk of Mortality. Dissertation, University of Michigan.

233. Roberts, I. 1996. Smoke Alarm Use: Prevalence and Household Predictors. *Injury Prevention* 2(4): 263–265.

234. Smolinske, S. C. and Kaufmann, M. 2005. Consumer Perception of Household Hazardous Waste. *Journal of Toxicology &. Clinical Toxicology.* 43(6): 706–707.

235. Wald, B. G. and Tope, P. H. 1971. Household Noise Problems—Appliances. *Journal of the Acoustical Society of America* 49(1): 117.

236. Warner, B. L., et al. 2003. Washing Machine Related Injuries in Children: A Continuing Threat. *Injury Prevention* 9(4): 357–360.

237. Wiley, C. C. and Casey, R. 1993. Family Experiences, Attitudes, and Household Safety Practices Regarding Firearms. *Clinical Pediatrics* 32(2): 71–76.

Sector 92: Public Administration

921: Executive, Legislative, and Other General Government Support
922: Justice, Public Order, and Safety Activities
923: Administration of Human Resource Programs
924: Administration of Environmental Quality Programs
925: Administration of Housing Programs, Urban Planning, and Community Development
926: Administration of Economic Programs
927: Space Research and Technology
928: National Security and International Affairs

READINGS IN HISTORY

1. Aday, R. H. 2003. *Aging Prisoners: Crisis in American Corrections.* Westport, CT: Greenwood Press.

2. Aiken, M. and Alford, R. 1970. Community Structure and Innovation: The Case of Public Housing. *The American Political Science Review* 64(3): 843–864.

3. Anders, G. C. 1986. Incompatible Goals in Unconventional Organization: The Politics of Alaska Native Corporations. *Organization Studies* 7(3): 213–233.

4. Andrew, W. L., et al. 1999. The Deep Structure of Presidential Space. *The American Review of Public Administration.* 29(3): 211–224.

5. Andrews, R. N. L. 1999. *Managing the Environment, Managing Ourselves: A History of American Environmental Policy.* New Haven, CT: Yale University Press.

6. Asch, P. 1988. *Consumer Safety Regulation: Putting a Price on Life and Limb*. New York: Oxford University Press.

7. Austin, J., et al. 2001. The Use of Incarceration in the United States. *Critical Criminology* 10(1): 1572–9877.

8. Banner, S. 2002. *The Death Penalty: An American History*. Cambridge, MA: Harvard University Press.

9. Bayless, G. J. 1895. *The Office of Mayor in the United States: A Study in Administrative Law*. New York: Columbia University.

10. Beyle, T. L. and Muchmore, L. R. 1983. *Being Governor: The View from the Office*. Durham, NC: Duke University Press.

11. Bilstein, R. E. 1989. *Orders of Magnitude: A History of the NACA and NASA 1915–1990* (NASA SP-4406). Washington, DC: NASA.

12. Bishop, C. F. 1968. *History of Elections in the American Colonies*. Manchester, NH: Ayer Publishing.

13. Blomberg, T. G. and Lucken, K. 2000. *American Penology: A History of Control*. Piscataway, NJ: Aldine Transaction.

14. Bottomley, A. K. 1990. Parole in Transition: A Comparative Study of Origins, Developments, and Prospects for the 1990s. *Crime and Justice* 12: 319–374.

15. Brady, D., et al. 1989. Heterogeneous Parties and Political Organization: The U.S. Senate, 1880–1920. *Legislative Studies Quarterly* 14(2): 205–223.

16. Bromberg, J. L. 2000. *NASA and the Space Industry*. Baltimore, MD: Johns Hopkins University Press.

17. Broome, B. J. 1995. Collective Design of the Future: Structural Analysis of Tribal Vision Statements. *American Indian Quarterly* 19(2): 205–227.

18. Declercq, E. and Kaminski, J. 1978. A New Look at the Office of Lieutenant Governor. *Public Administration Review* 38(3): 256–261.

19. Godschalk, D. R. 1999. *Natural Hazard Mitigation: Recasting Disaster Policy and Planning*. Washington, DC: Island Press.

20. Hall, P. G. 2002. *Cities of Tomorrow: An Intellectual History of Urban Planning and Design in the 20th Century*. Boston, MA: Blackwell Publishing.

21. Hays, R. A. 1985. *The Federal Government and Urban Housing: Ideology and Change in Public Policy*. Albany: State University of New York Press.

22. Hickman, H. L. and Eldredge, R. W. 2001. A Brief History of Solid Waste Management in the U.S. During the Last 50 Years. In *Municipal Solid Waste Management*.

23. Holli, M. G. 1999. *The American Mayor: The Best & Worst Big City Leaders*. University Park, PA: Penn State Press.

24. Hood, R. C. 1954. Reorganizing the Council of Economic Advisers. *Political Science Quarterly* 69(3): 413–437.

25. Hoogenboom, A. 1959. The Pendleton Act and the Civil Service. *The American Historical Review* 64(2): 301–318.

26. Irwin, J. 1985. *The Jail: Managing the Underclass in American Society*. Berkeley, CA: University of California Press.

27. Johnson, D. R. 1981. *American Law Enforcement: A History*. St. Louis: Forum Press.

28. Klein, D. E. 2002. *Making Law in the United States Courts of Appeal*. Cambridge: Cambridge University Press.

29. Lewis, J. 1985. The Birth of the EPA. *EPA Journal* (November). http://www.epa.gov/history/topics/epa/15c.htm (accessed 20 November 2008).

30. Lewis, O. F. 2005. *The Development of American Prisons and Prison Customs 1776–1845*. Whitefish, MT: Kessinger Publishing.

31. McBeath, G. A. and Morehouse, T. A. 1980. *The Dynamics of Alaska Native Self-Government*. Lanham, MD: University Press of America.

32. McDiarmid, J. 1946. The Changing Role of the U.S. Civil Service Commission. *The American Political Science Review* 40(6): 1067–1096.

33. McLoughlin, D. 1985. A Framework for Integrated Emergency Management. *Public Administrating Review* 45: 165–172.

34. Merryman, J. 1987. Civil Law Tradition. *The American Journal of Comparative Law*. 35(2): 438–441.

35. Messenger, S. L., et al. 1985. The Foundations of Parole in California. *Law & Society Review* 19(1): 69–106.

36. Mintz, J. A. 1995. *Enforcement at the EPA: High Stakes and Hard Choices*. Austin, TX: University of Texas Press.

37. Monkkonen E. H. 1992. History of Urban Police. *Crime and Justice* 15: 547–580.

38. ———. 2004. *Police in Urban America, 1860–1920*. Cambridge: Cambridge University Press.

39. Newton, J. 1997. Federal Legislation for Disaster Mitigation: A Comparative Assessment between Canada and the United States. *Natural Hazards* 16(2–3): 219–241.

40. Payson, S. 1994. *Using Historical Information to Identify Consumer Concerns about Food Safety*. Washington, DC: U.S. Department of Agriculture, Economic Research Service.

41. Pollock, J. M. 2006. *Prisons: Today and Tomorrow*. Sudbury, MA: Jones and Bartlett Publishers.

42. Polsby, N. W. 1968. The Institutionalization of the U.S. House of Representatives. *The American Political Science Review* 62(1): 144–168.

43. Pommersheim, F. 1995. *Braid of Feathers: American Indian Law and Contemporary Tribal Life*. Berkeley, CA: University of California Press.

44. Relyea, H. 1997. *The Executive Office of the President: A Historical, Biographical and Bibliographical Guide*. Westport, CT: Greenwood Press.

45. Reps, J. W. 1992. *The Making of Urban America: A History of City Planning in the United States.* Princeton, NJ: Princeton University Press.

46. Rosen, G. 1993. *A History of Public Health*. Baltimore, MD: Johns Hopkins University Press.

47. Rosholt, R. L. 1966. *An Administrative History of NASA, 1958–1963*. U.S. Government Printing Office.

48. Sellin, T. 1931. The Historical Background of Our Prisons. *Annals of the American Academy of Political and Social Science* 157: 1–5.

49. Sinclair, B. 1989. *The Transformation of the U.S. Senate*. Baltimore, MD: Johns Hopkins University Press.

50. Squire, P. 1992. Legislative Professionalization and Membership Diversity in State Legislatures. *Legislative Studies Quarterly* 17(1): 69–79.

51. Thacker, S. B., et al. 1988. Public Health Surveillance in the United States. *Epidemiologic Reviews* 10(1): 164–190.

52. Tonry, M., and Petersilia, J. 1999. American Prisons at the Beginning of the 21st Century. *Crime and Justice* 26: 1–16.

53. Uchida, C. D. 2001. The Development of the American Police: An Historical Overview. In *Critical Issues in Policing*. Prospect Heights, IL: Waveland.

54. Vanstone, M. 2004. *Supervising Offenders in the Community: A History of Probation Theory and Practice*. Farnham, UK: Ashgate Publishing.

55. Villa, B. and Morris, C. 1999. *The Role of Police in American Society: A Documentary History*. Westport, CT: Greenwood Press.

56. Warren, C. 1922. *The Supreme Court in United States History*. New York: Little Brown.

57. Watkins, S. 2000. Developing Statewide Emergency and Disaster Preparedness Expertise. *Journal of the American Institute for Conservation* 39(1): 165–172.

58. Welsh, M. E. 1970. The Role of the Governor in the 1970s. *Public Administration Review* 30(1): 24–26.

59. Wilkins, D. E. 2006. *American Indian Politics and the American Political System*. Lanham, MD: Rowman & Littlefield.

60. Williams, V. L. 1996. *Dictionary of American Penology*. Westport, CT: Greenwood Press.

61. Wines, F. H. 1895. *Punishment and Reformation: An Historical Sketch of the Rise of the Penitentiary System*. New York: T. Y. Crowell.

Geographic Index

Page numbers are listed first; entry numbers follow inside parentheses.

Industry Index

Page numbers are listed first; entry numbers follow inside parentheses.

Company Index

Page numbers are listed first; entry numbers follow inside parentheses.

About the Author

Dale Stirling has enjoyed a wide-ranging career that began in 1980 as a historian with the Alaska Department of Natural Resources. During six years with that agency, he became an expert in the history of Alaskan rural transportation systems. Between 1987 and 1996, he worked as an environmental historian with several prominent environmental engineering firms located in the Seattle metropolitan region. He became nationally known as an expert in environmental due diligence and served on a related advisory board in the 1990s with the Washington Department of Ecology. In 1998, he applied his historical research skills to the public and environmental health field while working at a toxicology consulting firm in Seattle. Since 2005, he has worked as a sole-source consultant on environmental and public and environmental health issues for a wide variety of clients, including law firms, insurance companies, and large engineering and environmental consulting firms, as well as private corporations and individuals.

Stirling has published extensively during his career, including books, peer-reviewed journal articles, monographs, reports, and hundreds of book reviews for scholarly journals. He has also served on numerous advisory boards and is a keen proponent of friends-of-library organizations.

In 2007, after 20 years of living in the Seattle area, Stirling and his wife Stephanie moved to a quiet five-acre parcel on Whidbey Island. They now enjoy deer, rabbits, eagles, and other wildlife over the sounds of Interstate 5 and never-ending construction.

www.ingramcontent.com/pod-product-compliance
Ingram Content Group UK Ltd.
Pitfield, Milton Keynes, MK11 3LW, UK
UKHW040909190926
475560UK00008B/27

9 780810 867017